alibris

KISSELBURG MILITARY BOOKS
7824 HIDDEN MEADOW
TERRACE
POTOMAC, MD 20854
UNITED STATES

To: ALIBRIS APEX DC 76670941-86
APEX
800 AVONDALE AVE.
GRANDVIEW HEIGHTS, OH 43212-3473

- -

Shipping Instructions for KMBOOKSS

Print this Packing Slip and enclose inside the front cover of the book.

Please ship this item no later than Fri Aug 14, 2026.

Ship to:

ALIBRIS APEX DC 76670941-86
APEX
800 AVONDALE AVE.
GRANDVIEW HEIGHTS, OH 43212-3473
UNITED STATES

"This Is A Good Book!"

In spite of my library school training in book reviewing never to say "This is a good book;" "This **is** a good book," is the best description I can give for *Combat Surgeon*!

The author's and my paths must have first crossed in 1930-1931, when I was a freshman at Wofford College. In 1934, Dr. Herbert succeeded me as editor of *The Old Gold and Black*, our student newspaper at Wofford.

Combat Surgeon, written for his family members, will also attract other readers who will be impressed by accounts of his life on the Wofford College campus, Spartanburg, SC, medical training at Vanderbilt University and his service in the U.S. Army Medical Corps before and during World War II.

We met in a Metro station in Paris in 1945 when I looked up the stairway to see Major Herbert coming down. This seems as if only yesterday. I was delighted to see him. Our paths did not cross again during the war. Often I passed by the Hotel George Cinq, but being only a Captain, I could not share its goodies with him, which were for Majors and above! Both of us were separated from active duty at Fort Bragg, N.C. in late 1945. He was one of the very few friends or relatives I met overseas during World War II.

I read the manuscript in two sittings, eyes glued to every page.

"Never a dull moment" describes my reaction to his story, which I believe his family, veterans, friends and the Wofford College community will enjoy as much as I did.

Herbert Hucks, Jr.
Archivist, Wofford College

William Chapman Herbert, Jr. M.D.

COMBAT SURGEON

World War II Memoirs of a
U.S. Army Medical Corps Surgeon

By William C. Herbert, Jr., M.D.

Honoribus Press
Spartanburg, South Carolina

Published by
THE HONORIBUS PRESS
POST OFFICE BOX 4872
SPARTANBURG, SC 29305

An HONORIBUS PRESS
WAR NON-FICTION BOOK

ISBN: 0-9622166-6-6

Printed in the United States of America.

Altman Printing Co., Inc.

DEDICATION

These writings are lovingly dedicated to the innumerable wives and mothers, who with inner grief and outer pride, have sent their husbands, lovers and sons to unknown places over the globe, urging them to combat the evil forces which are among us even to this day. Especially do I dedicate this to my Loving Wife, my Infallible Critic, and Patient Counselor, who has indeed borne all of the above during the past fifty years.

William C. Herbert, Jr., M.D.

Morning toilette in a Field Hosptial

Author - Face herbage courtesy of General Patton and no hot water.

ACKNOWLEDGEMENTS

These writings would not have been forthcoming had it not been for the unrelenting and persistent urgings of my three children: Bill, the eldest who should be writing his own stories, for he is par excellence, a superb story teller, my daughter "Bev" and the youngest John; my "Infallible Critic," and supporter, Gale, also bears much of the credit (or blame) for many of the following stories, for she is forever extolling me to "write it down." So there, it is her fault that all these embroidered tales she has endured hearing so oft repeated are now in print.

And again my debt to Wofford College grows, for I am indeed grateful to Doctor Dan Olds, Director of Computer Services. His invaluable assistance along with the renewed opening of the extended world of word processing by his able computer lab assistant, Watts Hudgens, has made this task much more enjoyable.

My appreciation is extended also to Doctor Lynn Fredrickson, formerly "Captain Fredrickson" and the Reunions of the 48th General Hospital for supplying me with many stories which we have "swapped" when we have met over the years. The tale tellers have "never spoiled a story for lack of facts." Captain Fredrickson shared our wild ride across France with the 54th Field Hospital following General Patton, as did several of the nurses and enlisted men.

I am also grateful for the help extended by Doctor Lewis Jones, Mr. Herbert Hucks, Ms. Cathy Rainey and Doctor Charles Webb in the completion of these stories.

Soon after my friend Doctor Sam Fleming brought out his book detailing his adventures in World War II, a volume entitled *FLYING WITH THE HELL'S ANGELS*, I chanced to be talking with Colonel Ed Hall, Sam's co-author, at Wofford College and mentioned that I had some stories as funny as Sam's. His immediate response was "write it down, and bring it to me." So began this rich and rewarding help in producing this happy tome. Without his assistance and encouragement it would not have been produced.

Happy Wedding Couple
Oklahoma City
27 August 1941

TABLE OF CONTENTS

Dedication
Acknowledgements
Prologue

PART ONE
Early Years 13

PART TWO
Wofford College Student,
Vanderbilt Medical School, and Off to Prison 45

PART THREE
The War Years 79

Epilogue 149

The Three Mousequeteers return to the 48th General in Paris from the battlefield.

PROLOGUE

"If you do not know what writing is, you may think it is not especially difficult. Let me tell you that it is an arduous task: it destroys your eyesight, bends your spine, squeezes your stomach and your sides, pinches your lower back and makes your whole body ache. Like the sailor arriving at the port, so the writer rejoices on arriving at the last line. De gratias semper." Colophon of a 12th century Beatus manuscript.

Centuries ago, one of my ancestors made apologies for writing about himself. I make no such, for the whole of the research and writing thus far has been enjoyable. I plan to be even happier in writing about something of which I know from first hand knowledge, i.e., myself. Of course the tales will be biased and prejudiced, for as Gale, (my Infallible Critic) has mentioned, I "never spoil a story for lack of facts."

Also, I have long wanted for the children and perhaps my grandchildren to know and understand what made our personalities what they are. I am sure they never knew how close to the mark they have been when they imply that we have "depression mentalities." Only those who lived through the Thirties and the subsequent World War can understand the forces which shaped our lives.

The following pages are devoted to, in so far as my powers of description are able to convey the trials, tribulations, joys, triumphs, and overall happiness this child of this century. If I do not succeed, so be it. I will have tried and had fun trying. Long live the computer.

How many times has an aspiring author been asked the question. "What motivated you to begin writing something like this in the first place?" As I ask myself the same question, I realize that many factors have played a part. The most outstanding reason has been the insistent pressure over the years that my children have applied saying, "You have to write down these tales."

Who knows what honors and blame will be laid upon the use of the Personal Computer in future time, for the avalanche of writings pouring out upon an unsuspecting public. The computer is in many ways comparable to Woman. There is no end to the secrets yet to be found by the uninitiated computer novice. Many of the mysteries cannot be unlocked, but the infinite fascination, as it is with the female, continues to be insoluble. Who could have foreseen that as poor a speller as I would only have to punch in "Spellcheck" in order to make me a perfectionist in spelling. I, who a short while ago agreed with Thomas Jefferson that correct spelling was whatever I wanted it to be. (That erudite individual spelled one word five different ways in just one letter.) Now I am driven by the malignant machine I face to punch in "Spellcheck" or "Dictionary" or "Thesaurus" before I am able to turn off the beast. It is only necessary to punch the proper buttons and this page comes out of the printer in Egyptian Hieroglyphics, Greek, Latin or any other language fed into the bowels of the monster. Of course, I have no earthly idea what the writing means as it spews out the maw of the beast. "WHAT HATH MAN WROUGHT?"

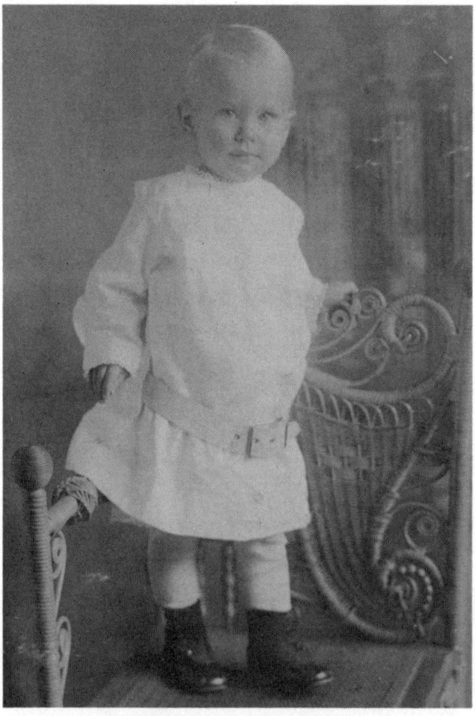

William C. Herbert, Jr.
Age two years - 1916

PART ONE

EARLY YEARS

It is very difficult, if not impossible, to differentiate "the Wofford Years" from the rest of my life because the influence of Wofford has permeated it from the beginning. It still does. My grandfather Walter Isaac Herbert (1858-1930), attended Wofford in 1873. This fact cannot be proved, because the "Damn Yankees" (I was in college before I learned that this was two words instead of one) in the form of soldiers, imposed upon us during the so called Reconstruction years destroyed the records in that era. (How the grinding poverty, the humiliation of a military occupation and the taxation of a Negro government dictated by Carpet baggers from the North could possibly be called Reconstruction escapes comprehension?) The notes issued to pay for the gold railings and the gold spittoons in the South Carolina State House were retired only in the past twenty years. I have checked with Mr. Herbert Hucks repeatedly whenever I thought I had a clue as to my grandfather's Wofford Years with no beneficial result.

My Aunt Helen, his daughter who is now ninety-five and very alert recalls that he often bragged that during his stay at Wofford he had the best looking boots in the school. This was important in that the footwear was made on the family plantation, undoubtedly by an ex-slave. I have letters from two Confederate soldiers requesting boots to be made and sent to them in the Army of Virginia. There is also a series of letters from my ancestress who was attending the Spartanburg Female College at the time, stating that she had seen her brother and was allowed to nod to him as he sat across the sanctuary of the Methodist Church. How long he went to Wofford is not recorded.

At any rate the Wofford connection was all pervasive in my family. Note the postcard with the picture of Doctor James H. Carlisle, President of Wofford, as the feature of the postcard. It was saved in my grandmother's memorabilia. The handwriting describing the great "Doctor" as "the Gladstone of South Carolina – the character builder of Wofford" is in my father's Spencerian penmanship.

At last count, which was recently, there have been some thirty of the Herbert name to attend the renowned school. At the present time, my grandson bears the burden of being reminded that numerous Herberts have preceded him at Wofford.

When my father was eighteen, his father determined that he should have a college education. Until that time he had attended a one-room school near the family home in Newberry County, close to the crossroads of Deadfall. The schoolmaster was Mr. Thornwell Hughes, who was subsequently the United States Ambassador to the Court of St. James, England. There were thirty- two students in the school at the time the picture was taken around 1896. There were no disciplinary problems, according to my Aunt Helen's recollection.

My grandfather and father boarded a train in Newberry to get to Union, S.C., in 1900. At that station they secured a wagon to bring them on to the college, along with the new trunk Dad had purchased for the great adventure. When they arrived at Old Main, the prospective student mixed with the students who were signing up for the various classes. "Don't take the Latin or the Greek" was one suggestion. "Don't take Mathematics under any circumstances" was another. In a short while, his father appeared, saying, "Let's go get some lunch. I'm getting hungry." Dad replied, "I haven't signed up for all my classes yet." "Oh, yes you have" was his father's answer. "You are entered in the Greek, Latin, and Mathematics classes as well as in English and Philosophy." There was no recourse.

These served Dad well in that he taught Greek, Latin and higher Mathematics for many years. There were about five professors during my time in Wofford who could respond when the call came that Doctor Rembert or Professor Gamewell was ill and a class in Greek, Latin, Calculus or other studies needed a substitute. Those were the days of real scholars. I have heard that Doctor Snyder, in order to obtain his degree at Heidelberg University in Germany, was required to translate Chaucer into Greek, then into Latin, and finally into the German of the Middle Ages as his final exercise for graduating.

I was brought to Wofford whether or not I wished it in 1918. I had no choice since I was age four at the time. My father had been selected to be Headmaster of The Wofford Fitting School, a position which he was to fill until it merged with the main college. At the time, it was a military school during the Great War (1914-1918), "the War to End All Wars." Discipline was strict. Dad was the right man for the job, for he was a strict disciplinarian, both on himself and others.

He was also the purchasing agent for the school. During one purchase, he found a great bargain in olives for the cafeteria. He forgot that the boys were mostly from the farms and had never tasted the eastern delicacy. Most of them rejected it, and the teachers ate olives until the large keg was empty, a long, long time later.

The classes in the fitting school were taught by my father, Professor A. Mason DuPre, a previous headmaster, and Mr. L. A. Odom, who became a lawyer of note in the community. On one occasion, Mr. Odom had a date whom he wished to impress. Dad had just acquired a brand new uniform, Sam Brown Belt, riding breeches, and gleaming new boots: all the equipment of the army officer of 1918. He let Mr. Odom borrow the new uniform for the evening, and thereby made a lifelong friend.

Mr. Odom eventually took his legal degree from Clemson, probably in the last class to obtain a law degree from that institution. In later years he acquired a pony in some trade, and since he had no place to keep it, we, the Herbert children, had the dubious pleasure of grooming it, feeding it and sometimes even riding it. The large pony had learned the trick of putting her head down between her legs and dumping the unsuspecting rider, along with the saddle, over her head. In other words, we could ride as long as it suited the little beast. Even Dad could not pull the cinch tight enough to prevent a dumping. This factor considerably curtailed the joy of owning a pony. A valuable lesson was learned. The way to cure a child of the desire to own a pony is to let the child have the pleasure of feeding, cleaning, grooming, and clearing out the stable every day. We did have a good pile of manure that year to put on the plants and garden.

At the age of four, I recall reading the Spartanburg Herald. The headline is vivid in my memory "WAR ENDS." It was a time of great celebration for not only did the War end, but the deadly plague of "Influenza" ended at the same time. The quarantine, previously imposed on the whole state, was lifted on November 11, 1918. The Armistice was announced the same date. Consequently, people could once again go to church, to school, to the pool hall, to the picture show, and the housewives were finally allowed to sweep their houses again. All congregating had been forbidden, because this disease killed more people in three months than all the plagues of recorded history: some twenty million world wide, with three hundred from Spartanburg.

The Mayor of Spartanburg was alerted that the war was over by the "new fangled device:" the Telephone. Of course this was confirmed by the old reliable Western Union telegram. The Mayor alerted everybody in town by preempting a railroad engine and riding it up and down the track which crossed the center of Spartanburg, blowing the whistle and ringing the bell. The church bells joined in. Bonfires were lit, and everywhere there was dancing in the streets (most of them unpaved). The young boys in Wofford Fitting School wanted to join in, but the stern announcement that any boy who left campus would be shipped home immediately curbed the insurrection,. This order was compensated by an announcement that there would be a parade the next day, and all three companies of cadets would take part along with a contingent from the nearby Camp Wadsworth. Except for this marching group, the soldiers also were confined to the camp and refused permission to enter Spartanburg probably because of the "flu."

Growing up on Wofford Campus gave me the privilege of knowing many of the professors. Doctor Lewis Jones, the esteemed historian at Wofford, now semi-retired, and I were in Old Main some years ago gazing at the pictures of the professors. Upon comparing notes, we found that he had known personally all the teachers except the first nine, whereas I had been acquainted with all but the first seven. This gives some idea of the longevity of the Wofford greats, for

Doctor James H. Carlisle was one of the signers of the original Article of Secession, a copy of which is in the Spartanburg Museum. (One of the original documents is in the Museum in Union.) Wofford opened in 1854, and the "War of Northern Aggression" also known in later years by the more genteel ladies of the South as "the Late Unpleasantness between the States" began in 1861. Doctor Carlisle was on a par with General Robert E. Lee when evaluated by my father and his friends in their stories of Wofford. As a student Dad had known and revered Doctor Carlisle.

One of my favorites was Professor Rembert, whom my father said was the most brilliant teacher on the Wofford Faculty. Professor Rembert and I became friends when he described to me as a young lad how a mule came to drown in front of his home. "Knotty" (a student nickname) Rembert lived on North Church, on the corner of what is now Daniel Morgan Avenue. Across from his home on the edge of the road was a spring which usually maintained a large mud puddle. Once a wagon loaded with cotton bales turned over at the edge of the puddle, plunging the mule into the water to drown before he could be freed of the load. North Church Street was unpaved during my boyhood. My feet still feel good when I remember the soft dust or the cool mud coming up between my toes as I walked to town. I also recall how hot the railroad bridge could get on the bare feet. (We all went barefooted in the summer.) Doctor George Dean Johnson recalls how, as school children, each year before entering school, we had to bring a stool specimen to the "Hookworm Hospital" (really named the Good Samaritan Hospital). I also remember being held down by my father and Doctor Sanders, while the Doctor swabbed out a hole in my heel with a cue tip dipped in an iodine solution. This ostensibly was used to kill the Tetanus germs which may have been inserted when I stepped on a rusty nail. (Anti-tetanus serum had not been invented at the time.) As an added precaution the wound was also packed open with iodoform gauze. This gauze was removed only a quarter inch a day. (What torture and what poor medicine.)

Doctor Rembert became an even better friend when I found that he was as much a devotee of mystery novels as I was. It was a strange friendship indeed; that of a callow teenager and the "most brilliant" man at Wofford. When I knew him he was a wizened old man with a brain pan twice as large as his small statue should carry. He made a vivid picture, his sparse unkempt hair flying in all directions as he talked on and on, mixing Greek Mythology and modern detective tales. When he died, he left me his collection of mystery books.

Incidentally, the first book I could claim as my very own was a childhood version of Greek Mythology. To those of you not well acquainted with the Greek tales, those stories make the Grimms Fairy tales pale in comparison. If you want to read sex and violence go back to the Greeks. My favorite was the story of the journey of Theseus to Athens where he encounters the hospitality of Procrustes, the Giant, who offered the traveler a bed like no other. If the sleeper was too long for the bed his feet were cut off to fit, and if the person was too short, the unsuspecting guest was stretched to fit. No wonder I still recall my first book with such relish.

Doctor Clinkscales "Clink" (as he was known to us) still lives in the memory of my generation of students. He still inhabits the classroom on the second floor. His first instruction to me was "Don't sit there, Bill. Sit in that place by the window. That's where your father sat when he came to this class." Even thirty years later he would greet a former student by his first name as if he had seen him only the day before.

On the first day of a new class, he read the roll. "Sidenstricker. That's your name?" "Yes, Sir." replied the victim. "Son, that [harrumph] ain't a name, that's a disease."

Who of our class can ever forget the gracious and flowing speeches of Doctor Henry Nelson Snyder. He was said to be one of the South's greatest orators. He was asked to speak on one occasion, and he asked how long he should talk. The reason was that if he were asked to talk thirty minutes or an hour, he could begin immediately, but a five-minute talk required at least a month's preparation. When Doctor Snyder gave the sermon in Central Church, one of my tasks was to find all the extra chairs in the building and place them in the sanctuary. On one occasion, our Wofford class had been guilty of unsportsmanlike conduct at a recent football game. Chapel was called the next day, and Doctor Snyder verbally chastised us so thoroughly that we slunk out of the building like "Whipped Curs."

Nobody could mistake that erect figure moving briskly toward his home among the trees. It could only be David "Dunk" Wallace. Even today his scholarship remains unchallenged. When I was a lad, I made a golf club by cutting a limb off a tree, the smaller portion was the shaft, and the larger was fashioned into the head. I was hitting a tennis ball around the campus one Sunday, when he called my father to stop the game of golf on Sunday. ("'Tis said there warn't no card playing, no dancing, and no drinkin' on Wofford Campus in them days!")

On one occasion, Doctor Wallace was called to task by a faculty member because of his poor handwriting. He replied that though his handwriting was not pretty it was very legible. To which my father replied, "Doctor Wallace, your handwriting is so poor I can't even read your typing." Dad's handwriting was so beautiful that he made spending money during his college days by writing visiting cards.

I would not omit my father's name from this distinguished company. The "Colonel" long walked the Wofford path gaining the accolade from his colleagues as "the finest of Christian Gentlemen." One of the Wofford coaches once said he would not dare make his teams go through the physical disciplines that Dad required of himself every day.

"Fesser Coates," (Professor K. D. Coates) the gentlest of teachers, should be remembered also. On the first day of class, he announced that in order to encourage parallel reading, he would accept anything upon which we chose to report. "I can't believe this guy" was my private reaction. In order to plague him, I turned in *Tarzan of the Apes.* He laughed and accepted the report. I give him many thanks for his understanding that *Tarzan of the Apes* is needed to leaven *Paradise Lost* for the average freshman. Eventually we became the best of friends. He was only five feet six inches high, but was coach for the tennis team. Later I found out that he bought the tennis balls for the matches out of his own meager salary and that during the depression!

"Fesser Coates" once told this tale on himself. He met an old Grad on the campus one afternoon, and couldn't place his name. The alumnus said "Aw, Fesser Coates, you remember me. You larned me English."

Doctor Chiles is recalled as a rather gruff, heavyset, aloof individual who had a deep chuckle when amused. He wrote several texts for teaching German.

In the "good ole days" a student could take courses for fun and I succeeded in taking Surveying and Electricity under "Peg Shuler" for the sheer joy of knowing something unusual, though I have never used either in later life.

And let us not leave out "Fish" Salmon who helped many a wanderer with his cheerful guidance through the forest of irregular French verbs, nor can I resist mentioning "Peg" Bourne with an intellectual catharsis such as "You gentlemen don't know any German today, let's see if you know anything about automobiles."

I only knew Professor Gamewell when he was an old man. He had been a Confederate veteran, and was no longer active on the faculty as teacher of Greek and Latin. His main interest had been in the Wofford Lyceum. Once when I was plundering under the Cleveland Science Hall, I found a large box full of Magic Lantern slides, 3 x 4 inch negatives. These were used to illustrate lectures on Greek and Roman architecture. I wonder what ever became of them? I am sure these were part of Professor Gamewell's teaching tools.

"Cutie Norton," later to be Dean, will always be remembered for his annual Christmas Rendition of Scrouge. We have yet to see his equal in his rendering of *A Christmas Carol.*

Herbert Hucks recalls that one of Professor Pettis' habits was to talk to the class at the same time he erased the blackboard with his left hand. He called on Mr.F— to come to the board and write out the physics problem behind him. Mr F— was not one of the future scientists, and after writing out the problem, he was stuck. Meanwhile Professor Pettis continued to talk, and in a short while he began to erase the board. Abruptly, he stopped and apologized to Mr. F— for erasing his writing. "You did get the answer, didn't you?" was his query. "Yes, Sir" was the student's reply. The class had difficulty keeping from bursting out laughing.

Doctor Coleman Waller can best be remembered for his cigars. These were probably smoked to overlay the odors of the chemistry laboratory, which they certainly did. He would smoke the cigar for awhile and lay it aside on a window sill. There it would remain for an indefinite period gaining strength in its ability to override other odors. Doctor Waller's other notable contribution to our education was his habit of leaving his well thumbed Gray's *Anatomy* lying around. This helped satisfy the overwhelming curiosity as to the actual anatomy, both internal and external of those mysterious beings, i.e., girls.

"Rick" Patterson was probably the youngest of the teachers of my time. He took the work of preparing students for medical school quite seriously. He was noted for his "tough" courses: Chemistry V (Quantative Analysis) met the first day of classes and eleven students turned up. He announced that there was only room for five in the laboratory and after the six weeks test there would only be five in the class. (And so it proved.) I give him the credit for my being able to hold my own in the tougher chemistry courses to come at Vanderbilt in Medical School. Though we never knew our grades, our standing in the subjects was leaked out, and rumor had it that I was first in chemistry. This would not have been surprising, since I took all the chemistry courses Wofford offered, and was chemistry student assistant the last two years.

And thereby hangs another tale. Since Dad was a professor, his children were not required to pay tuition. But being lab assistant carried a small salary. Since these were the Depression Years, even a miniscule salary of $40 a semester was a worthwhile sum. I submitted the bill to Mr. J.K. Davis, the treasurer. He must have been selected by malignant fates to suffer the burdens of Wofford during the Great Depression. (At one time the college paid no salaries for the period of three months. Had it not been for the little grocery behind the college granting credit I'm sure we would have gone hungry many times.) He insisted that I did not deserve a salary since I paid no tuition. The argument was settled in a faculty meeting, and I did receive the pay.

One of the real characters of the old Wofford was "Doctor L__, Professor of Profanity." He was the general factum factorum. He was many times called to the President's house to see if he could restart the recalcitrant furnace which furnished the central heat of the old antebellum home. A maze of steam pipes and valves led off from the huge coal firebox, and shutting off the wrong one could have blown up the whole thing. After straightening out the valves and instructing Doctor Snyder in the care and feeding of the ancient heating system for the umpteenth time, he was heard to mutter, "You can't larn him nuthin."

On one occasion the ladies of the campus were in charge of the Annual Faculty Barbecue. The barbeque pits were down toward the railroad, near the present fraternity houses. Apparently it was my Mother's job to check on the cooking with "Doctor L—" doing the actual labor. She approached the pits and asked how the food was progressing. "Doctor L—" asked "Would you like to taste it?" It smelled delicious, and she was on the verge of accepting when she glanced at the cook's arms. The hands and forearms were pristine clean, whereas the upper arms were filthy dark. He had been expeditiously mixing the mixture by hand throughout the early cooking. She declined graciously.

This was in the times of the national experiment known as "Prohibition," and "Doctor L—" found it very difficult to obtain good liquor. He found it expedient to satisfy his thirst with shoe polish, canned heat, and other dubious sources of refreshment. The contents of these alternatives contained methyl alcohol, which might have given the same effect as ethyl alcohol but had the deleterious result of causing "gunbarrel" vision. In other words, his peripheral eyesight became more and more restricted, until he was looking down a tube of central vision. In order to determine the identity of the person talking to him, he had to put his face in close proximity to the other person. At the best of times, "Doctor L—" had little regard for sanitation. This morning he had the misfortune to encounter Mrs. Mason Dupre, the Dean's wife. He invaded what we now call the "inner aura" of the said lady to determine just whom she might be. After inhaling the cloud surrounding him, she reported the incident to the Dean, and shortly afterward Wofford lost one of its most memorable characters.

Speaking of alcohol, I don't recall there being a real problem at Wofford. There were isolated instances, of course, many examples of victims of rum and other addictions were seen on the streets occasionally. Reports of illicit stills being raided were often in the papers, but no alcohol was ever sanctioned on the campus. The closest I ever came to tasting it was when I was sixteen. I was ill for some reason, and Mrs. Warren Dupre, the church Organist, brought me some Wine Flavored Jello. Dad met her and talked with her, but the wine jello never made it to my room. It was poured out before I even got to smell it.

I made the statement that there was never any alcohol in our house, but Gale corrected me. On one occasion when she was living on the campus, the need arose for her to ask my mother if she had any medication for the relief of the old monthly affliction of women known as "the curse" or "ministerial problems." My mother went into the kitchen and after climbing on a kitchen stool, she reached behind a row of artichoke pickle jars, and behind a row of pickled peaches found a lone jar of dark liquid which proved to contain a highly potent relief of the problem of cramps. This unlabeled container proved to be Blackberry Cordial with a probable seventy percent alcohol content able to relieve the sufferings of any sort of abdominal pain.

One evening when I was a young lad, I was invited to dinner with the neighbors, the Trawicks. Much to my surprise, when I arrived with my father, the other guests were several of the Wofford Faculty. Doctor Snyder was there, as was Professor DuPre, and four or five other of the older members. None of the wives were invited. After a delicious dinner, all the menfolk adjourned to the parlor (French - Le Parley meaning - to talk) and they did talk until eleven or twelve o'clock, well past my bedtime. They exchanged wonderful stories. How I wish a recorder had been available then. Never was a shady story recounted, nor were any four letter words used.

One of my long-time friends was a black man named Posey. He could have been the original of the cartoon character named Hambone, who was famous for his pithy comments and wise sayings. His head was as bald as an onion and running across it from ear to ear was a deep sulcus dividing the two portions of his brain pan into separate lobes. He was a dear man, always smiling, always ready to help, and he was in charge of the "Woop" wagon, later the "Woop" truck. In other words, he and a helper were charged with picking up the debris, tree limbs, leaves and what-ever litter might accumulate over the acres of the campus. When he wanted the wagon or the truck to stop, he loudly said "Woop" (whoa up - an expression from the cotton fields, and probably from slavery times).

Sunday afternoon entertainment for the children of the campus consisted of walking around the six-inch ledge of the Whiteford Smith Library, which is six feet above the ground in front but sixteen in the back. The alternative was to enter the library (all the professors had keys) and read the *Illustrated London News*.

The children of the campus were a rather disparate group in ages, but life under the "Twin Towers" did force us into a clan. Bob Wallace, son of the historian, was the eldest and a very energetic soul he was. His sister, Alice, was more our age, but generally backed up Bob in his escapades. Sophie, the eldest sister, was in college, and hence she had little influence on our activities. Doctor Snyder's grandson, Thomas Nelson was one of our playmates. His two sisters, Lula Dell and Frances, tried to join in but were somewhat young and "besides, they were girls." Carlos Moseley, grandson of Professor Dan Dupre, would join us occasionally, but he lived across town. Buck Thomson, son of Doctor Thomson, the dentist, was a member of the gang. Peggy Thomson Gignilliat was older and entirely wrapped up in becoming the finest violinist of the time (to graduate later from the Juliard Institute). She was also the most intelligent woman I have ever known. I once gave the Latin grace for the Medieval Dinner in honor of the high school students, and Peggy's remark was "You even got the case right." The Trawick twins and Helen Pugh were "on the fringes," but they "never quite made it" because they did not have bicycles.

All our activities centered around the bicycles. Bob had organized a club called the Pine Cone Bicycle Club with rules and regulations such as having to ride across a deep campus ditch bridged by a two by four plank. Riding a set distance with "no hands" was also in the requirements. The voice of the teenagers was also heard in the publishing of the weekly "Pine Cone Bicycle Club News," a one page back and front news letter which cost one penny. My duty was to decorate the heading with a stamp showing two pine cones, the branches and needles, and the name prominently displayed. The stamp which I made from the rubber of a discarded bicycle inner tube, was carefully cut out and glued to a piece of wood. How proud I was to produce such elegant and useful art work. The newspaper was a tremendous success, until Bob introduced a piece of campus gossip involving one of the faculty wives. Freedom of the press ceased to exist, and the paper was suppressed. No further issues were allowed to be printed.

All of the campus children were required to go to through a "survival course." It was named Fremont Elementary School, two blocks away. There the two sides of the tracks met, whether or no. The student body was composed of the children of Spartan Mills and Wofford Campus as a whole. Every recess, I had a standing appointment during the thirty minutes for a fight with a fellow named Sanford Wolfe. After fighting for some months, we became pretty good friends. I met Sanford some years back. He was about six feet in height and must have weighed 250 pounds, well proportioned. I was glad he no longer chose to fight.

Another memory of the children of Fremont School remains. During the winter, an attempt was made by the parent's of the younger children to prevent the ubiquitous colds and sniffles prevailing in the area. Many of the children wore a bag of asafoedita around their necks all winter to ward off the cold or "flu." To those of you unfamiliar with this folk remedy, I will need only say that the word derives from the Latin meaning fetid. The wearing of the bag was partially successful in warding off disease in that the wearer of the offending article was seldom allowed to associate with other humans, and therefore "germs" were not shared as indiscriminately.

The only teacher there that I recall was Miss Bessie Poole, who taught the third grade, and she is remembered not for her educational expertise but for her tolerance. At one period of quiet, she "snuck up" behind me. Instead of studying, I was drawing the back of the head of the fellow in front of me, on the blank page of the geography book. Horrors! (Today's children do not have this temptation, I'm told, for this subject is no longer taught in school.) Instead of a stern reprimand, she merely said, "Pretty good drawing," and passed on down the aisle. I still remember "Miss Bessie" with gratitude.

"Capt'n Jones," security for the campus, was also one of my friends though I wasn't sure of it at the time. Some repairs were being made at the intersection of North Church Street and Calhoun Street. (The Memorial Auditorium has covered this area now and the street has disappeared.) It once was an ideal tree-lined slope on which to risk life and limb on either skates or wagon. "Buck" Thomson and I found a treasure trove of hardened tar on North Church Street after the workmen left, and we struggled to lift a chunk into our wagon but could not. We then decided to roll the tar down hill then uphill to our hideout. Rolling it downhill was much easier than the task of rolling it uphill. About halfway up, we spied "Capt'n Jones" coming. We hid in the bushes. Instead of passing on by, he stopped, glanced at the tar, then looked around, and finally passed on down the hill. We rushed to rescue our ball of tar, but in the heat of the day, it had become stuck to the sidewalk. And there it stayed for many a year to remind the neighborhood children the evils of taking something which does not belong to them. The black stain perhaps is still there under the parking lot of the Memorial Auditorium.

One other encounter with "Cap'n Jones" comes to mind. The tennis courts were right in front of our house across the street, and were not maintained during the summer by the College. My brother Thad and I pulled up the weeds on the clay courts, and lined them. So naturally they "belonged to us." (The Trawick twins, Catherine and Louisa, would bring their dates up to play and suddenly we would rush to get there before them and to preempt the court. It still infuriates the Trawick girls when they recall our selfishness.) Mother knew not to serve the evening meal in summer until after it was too dark to see the balls.

I recall the first time that I saw the Trawick twins. I had been sent out to the woodpile to bring in some kindling to make a fire in the little hot water heater. At the time I had a hole in the seat of the pants I was wearing, which ordinarily did not concern me, but this time my father called me to meet our new neighbors, Professor and Mrs. Trawick accompanied by two identical visions of delight. Two girls were dressed attractively in pink and were about my age, and I had a hole on my breeches. I was as embarrassed as only a teenager can be, and carefully stayed backed up against the wall to hide my shameful condition.

But back to "Cap'n Jones." One summer morning the weather was just right for tennis, and as we began playing two lovely young girls with whom we were not acquainted arrived. They wore the short pleated tennis dresses which the new fangled Wimbledon codes had allowed. I was of the age when my "adolescent testosterone levels" were rising, and I found the visitors irresistible. We invited them to play. Shortly afterward, the nemesis arrived. "Capt'n Jones" called the two aside and after he talked to them, they departed. He made no explanation to us, but after discrete inquiries I found out that they were the "Joie de vie" girls who lived in the nearby Morgan Hotel on Magnolia Street. Come to think of it, how did "Capt'n Jones" know about their occupation?

For many years, Dad cut our hair with a hand clipper (not an electric one, which had not been invented). Frequently it had a habit of pulling and not cutting, so that getting a haircut was an ordeal, and the hair usually looked as if it had been cut with a bowl on top, but so did the haircuts of all my friends.

When I got a little older, only when absolutely necessary, I received a quarter to get a haircut on Saturday. What Dad did not know was that the Negro Barber College under the Cleveland Hotel would cut hair free just for the experience. The old Cleveland Hotel on Morgan Square has an enviable history. It was built on one of the two streets in Spartanburg paved with cobble stones. The stones have now been covered over, first by the brick road which Colonel Vanderbilt had had laid from the Hotel to Camp Wadsworth during World War One. (He took over the two upper floors of the Hotel as his residence for the family.) The road was paved with brick laid by the soldiers working after hours of training. The expense of the labor and the materials was borne by the Vanderbilts. The only other cobblestone paving, now visible in the town, is behind the warehouse where Mr. Wallace Dupre had his last auto store. (This is now where St. John's Street enters West Main Street.) The stream behind this area is called Nasty Branch with some reason, and the origin of the stream is the original spring in the center of town. It is now covered over by the parking lot in front of the Palmetto Bank. You can see and hear the stream running through a slave-built culvert under Harry's Restaurant which was formerly a second-hand furniture store named Jones' Furniture, owned by a distant relative of ours. I picked up two rusty andirons in the basement. After cleaning them I found them to be figures of George Washington, real treasures.

After going to the Negro Barber College, along with friends, came the exquisite pleasure of going to Woolworth's Five and Ten Cents Store. There one chose which of the stale, "on-sale" candies were the best value. These were usually caramels, and for 15 cents a large bag was filled. I don't care much for caramels to this day.

This left a dime to get into the Criterion Theatre. All the other movie houses charged 15 cents, and there was a good deal of warfare in the entertainment field. The Criterion was built near the area through which the railroad ran in the center of town. On one occasion someone put a bottle of Tincture of Valerium at the back of the Criterion. The board floor was sloped so the opened bottle made its way down to the front spewing its contents all over the floor. This liquid penetrated the wood so thoroughly that the floor had to be replaced. The bottle with the remaining liquid was brought to the

W. C. Herbert and A. C. Daniel, Jr.
Orators for Wofford Anniverary - 1904.

(Dad) Commandant, Wofford Fitting School

Herbert home on Calhoun Street.

Dr. James H. Carlisle
"The Gladstone of South Carolina.
The Character Builder of Wofford."

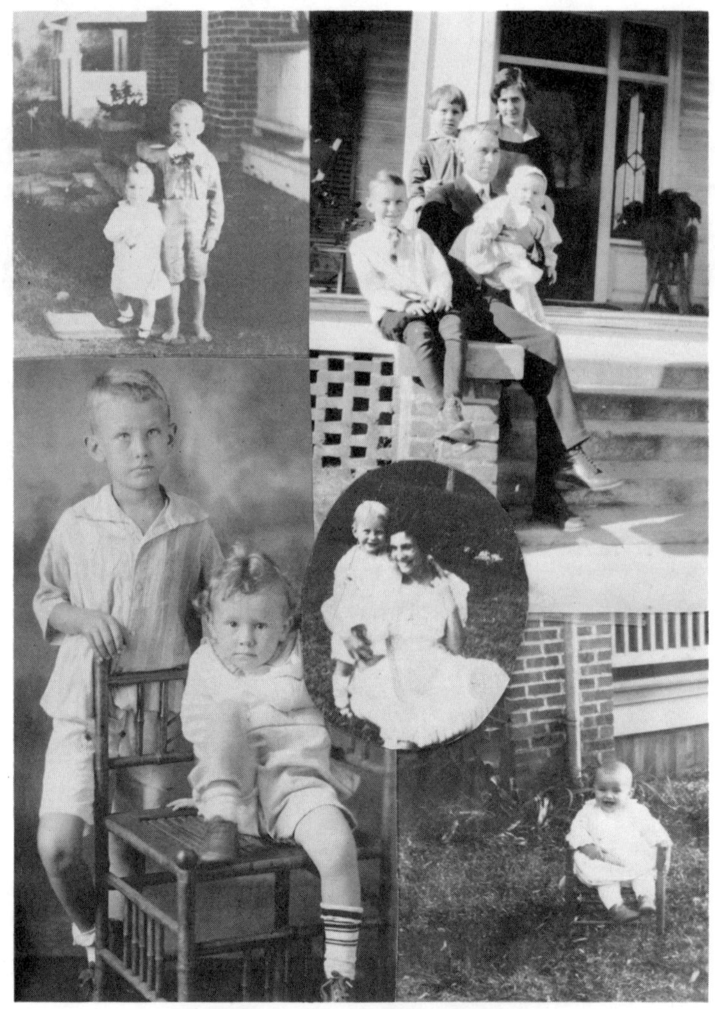

Herbert Family in the Twenties

The author, with intellectual companion.

Wofford "Old Main" - 1933

home of Doctor Waller, the Chemistry Professor for him to identify the chemical. The smell was so pervasive that I can recall the odor drifted across the Wofford Campus all the way from Doctor Waller's house (today the President's home) to our house on the corner of Calhoun Street.

As children we always wondered what was in the building across the street from the Criterion Theatre with the fascinating name, BURLESQUE, and no matter how many times we peered under the swinging doors we could never obtain a view of the spectacles presented. A few doors down the street was the magnificent theatre with the front done in Art Deco. This emporium had an organ which rose out of the floor in front of the stage with a spotlight on the "artiste." The organist turned out to be none other than our acquaintance, Tom Lyles. Tom was a few years older than our gang, and played the accompaniment for all the silent films. I especially remember the *Black Pirate*, with Douglas Fairbanks starring in the leading role.

The Saturday morning ritual at the Criterion included the world news, a feature film (usually a western) and was always concluded with an episode from the series of the Green Archer. (This was to assure that every child present would be back the next Saturday. Incidentally the Green Archer who righted all wrongs by shooting the evil doers with a green arrow, eventually turned out to be the town's kind gentle Doctor.)

Summers were long and leisurely. The only planned occupation I can recall was at Converse College. My parents noticed my drawings on occasions, and decided to send me to Doctor August Cook, the head of the Art Department at Converse and a well-known artist in his own right. The good doctor believed that the first essential of every artist was to learn to draw, so on entering the summer school class, I was told to sit down in front of a large plaster model of a crouching lion. With charcoal and paper, I was instructed to draw the beast. Toward the end of the morning session, Doctor Cook eyed the drawing, turned the lion halfway around and said "Draw it again." I continued to draw that "cotton picking" lion the whole summer. He was right, of course, but drawing that lion really became boring before the summer school ended.

One vacation was spent in the Hendersonville Laurel Park. The tiny hut had but one room, and it rained the entire week's vacation.

We were trapped, all five of us, in that little box, and I had only brought one book, *The Wizard of Oz.* What a boring week that was!

I never took a driver's test. When I reached the age that Dad thought appropriate (i.e., twelve), he asked me if I would like to learn to drive. He taught me. Carefully I'm sure, for I have never gotten a traffic ticket. The only warning I ever received was after eleven o'clock one night I was called to the Emergency Room at the Spartanburg General. After seeing the patient, I drove down Serpentine Drive to enter Wood Street. It was sleeting at the time, so instead of stopping at the Stop sign, I slowed and carefully went into Wood Street. Parked in plain sight in the empty lot where the Urologist Office now stands, was a highway patrol car. The officer flashed his light and gave his siren a slight touch. I pulled in beside him. "Doc, didn't you see me when you came up?" he asked. "No" I replied, "If I had I would have stopped!" "Well, I'll just give you a warning ticket" was his answer.

For some years I had the franchise to deliver the magazine *Saturday Evening Post* to the news stands the evening before the mail delivered the regular subscribers their copies. This involved driving all over downtown one afternoon a week and this precipitated the only wreck I ever had. As mentioned before, the railroad right of way ran across the center of Main Street about midway. At first there was no barrier, but after the fire engine was demolished and horses splattered all over that part of town by the inadvertent meeting of the fire engine and an outgoing train, the city fathers insisted that two gates be set to block the auto traffic around 1925. Considering that Spartanburg boasted only seven cars in 1907, this was no big deal. I was delivering the Post one afternoon, and the gate came down. There was one car ahead of me, and while the train was passing I watched a pretty girl beside me on the sidewalk. My adolescent pubes must have been rising at about that time. When the barrier rose, the car in front of me went forward, and I followed. He stopped however in a short distance, while I continued forward still keeping a sharp eye on the pretty girl swaying on my right. I hit the car without doing any damage, but I knocked the bumper off the front of the old Hudson I was driving. (I can still recall the humiliation of putting the bumper in the back seat.) Dad made little comment. The bumper was welded back in place without incident and probably at little cost. Come to think of it,

I never did tell my father the true story of how that bumper came to fall off the car. It really wouldn't have done any good and might have upset him. In fact the only father-son conversation we ever had was his giving me a slim book called *What a Young Boy Should Know*, and its companion slender volume *What a Young Man Should Know*. (Both books were put out by the Methodist Publishing House, and did not enjoy being on the best seller's list.) After reading them both, I did not acquire any additional knowledge that I had not already learned from more sophisticated schoolmates. My curiosity continues unabated to this day, for in spite of having been a gynecologist most of my life, I still find the human female the most fascinating, unpredictable, and rewarding riddle that any male can try to decipher.

Central Methodist Church has always played a large part in all the activities of Wofford College. This is not surprising in that Central was the first congregation organized in Spartanburg. (As per edict of Doctor Lewis Jones, Wofford Historian.) The baccalaureate ceremonies were in Central and usually an outstanding speaker was the center of attention. Extra chairs were put in the aisles by those of us big enough to carry a chair. The Herbert family was always at church, sitting in the fourth pew on the left side in the center. We remained there until Dad had his heart attack, induced by carbon monoxide in his bathroom which was heated by an unvented gas heater. We moved to the Sanhedrin (so named by Doctor Lewis Jones, the transept, on the eastern side) so we could get Dad out easily, or in case I got a beep. You have no idea how hard it is to get an unconscious 170 pound person out of a pew or movie theater seat until you have tried.

Central always had a delightfully cool sanctuary even in the hot summer time. It only came out lately that Dad would get up early on Sunday morning and go into the church. He would open the six big stained glass windows in each transept to allow the early morning breezes to sweep through the building, then close them before the morning service. Consequently we were greeted with a cool sanctuary even in the hottest weather. The Wofford professors were nearly all the Sunday School Teachers in those days.

On rare occasions I was allowed to turn the ratchet and pawl which was located to the left of the great middle window in the west transept. This allowed the great brass chandelier to be lowered in order to replace the light bulbs and polish the brass. This seem to have been Dad's adopted chore. These duties, in addition to teaching the boys class in Sunday School, seem to have fulfilled his ministerial ambitions at Central. He also served on the Boards, while Mother was always in the seat of the organ and directing the choir. Central has had only four organists since the instrument was installed in 1894, (at a cost of $4,000). These were Mrs. Warren Dupre, who began to play at age 16, followed by Mother, then Doctor John Bullard, and finally our present organist, Miss Beverly Barr. The music of Central has always been one of its outstanding ministries, and the quality has remained magnificent.

The town of Spartanburg was engulfed by religious revivals in the early days of this century, (though you couldn't tell it by the results). And they continued throughout our childhood. Across from Wofford, on the lot now occupied by the present county offices, and facing College Street was erected a large wooden revival building. Hundreds of people, carried along by the religious fervor, would attend at night. The children were required to go in the afternoon following school. I must say that no earthshaking experience graced my attendance, but I was tremendously impressed by Billy Sunday's Magic tricks. So much so that I attempted to become a magician. Perhaps this was not entirely due to Billy Sunday's efforts, for this was the era of Houdini, the famous magician and escape artist, my real hero. Nor can I really say that the frequent visits of the wooden tabernacle evangelists have improved the moral climate of the metropolis of Spartanburg, any more so than have the T.V. impresarios of the so called "religions" of today. My father once made the remark that the failure of Christianity was due to the fact that it has never really been tried.

Occasionally we went to the Baptist Church on East Main, a large rather hideous building built of yellow brick, and having an incongruous round tower in the front. Mrs. Pete Lyles, wife of the long time organist, Tom Lyles, of the Strand Theater fame, was in Singapore some years ago attending a religious meeting. One night at a large dinner she was seated next to a distinguished man, and the conversation revealed them both to be from the United States. She stated that her husband was the organist of the First Baptist Church in Spartanburg. She went on to comment what an awful architectural monstrosity the building was, only to be brought up short by the man next to her saying that he had been the architect of the said monstrosity.

We always had a garden behind the house, and usually a small chicken coop. It was my job to keep down the weeds, mow the lawn, clean out the chicken coop, and make sure there was kindling for starting a fire, both in the small hot water heater and sometimes in the large Warm Morning stove in the living room. Also I had to fill six coal scuttles every night during the winter months, even though the coal pile might be covered with snow. All the other rooms had fireplaces in which fires were seldom built as I recall. The large heater did a good job, but was filled with coal and consequently everything was covered with soot and ashes. But that didn't matter too much either, for the edge of the Campus bordered the railroad. According to Doctor Lewis Jones, a railroad enthusiast, several dozen trains a day came through Spartanburg, all belching sooty smoke. I recall playing rubber guns on the front porch, and ignoring the swirls of soot around me.

The fact that we were so close to the railroad did pose a problem, for this was during the Great Depression, and hordes of tramps rode the rails. It did not take them long to find out that our house was always good for a meal. Neither Dad, nor Mother, nor Grace, (the cook), would ever turn away anyone hungry. Miss Grace Donaldson, (daughter of Tom Donaldson the cook at Wofford's Carlisle Hall for many years), was a member of the family and still is. She joined our household when I was sixteen, and when my parents died she came to work for my family as long as she was active.

I remember one black lady who came up to the back porch one day and asked for food. While she was eating the meal that I had gotten from Grace, she asked my name. When I told her, she burst forth with "Why, Honey we's kinfolk!" It turned out that her name was Herbert also. Evidently her father had been a slave owned by the family, and had adopted the name upon being freed (a prevalent practice of the time).

Most of the Wofford Faculty members had automobiles in the early twenties. Doctor and Mrs. Pugh had an electric car, very much like a box with another box on top for the passengers and guided by a tiller in front of the driver. I recall it primarily because it was so quiet that it would creep up from behind without warning, whereas the other autos made a hearty satisfactory roar from the time they were stated by hand crank until the spark was killed. Sometimes, these maneuvers did not gentle the beast, for a loud backfire usually accompanied the attempt to stop the engine.

Early in the game, probably soon after WWI, my father went to the bank to borrow some money. The lending agent inquired what the purpose of the loan was to be. "I want to buy an Anderson automobile," was Dad's answer. "I'm sorry, Professor, but we can't lend money on such a risky investment, you understand."

We always had a car from the time I can remember, usually a big heavy Hudson. Dad did most of the upkeep work himself. I recall a trip when he took the entire transmission down on a hot dusty road in the country and replaced the worn leather cone of the clutch. One summer, he devoted his spare time to taking the engine apart, putting in new rings, grinding the valves, inserting new bearings for the drive shaft and pistons, installing new wirings and resetting the firing order of the six cylinders.

The only time I ever heard him use an expletive happened in connection with the car. Miss Lilly Strickland was visiting us. (This lady had been mother's roommate in Meredith College. Both of them were quite musical. Miss Strickland wrote extensivcly for the voice, among the selections was "*Honey, did you hear that Mockingbird sing last night?*") Dad hurried home from the office to take her to the train. He attempted to crank the motor but forgot to retard the spark. This frequently resulted in the crank handle flying around backward, breaking the cranker's arm. And that was what happened. All we ever heard Dad say was "My Stars, I've broken my arm." He always insisted that cursing was a sign that the fellow cursing lacked a sufficient vocabulary to express his feelings. The broken wrist was first set poorly by a local general practitioner, and Doctor Warren White, of the Shriner Hospital fame, reset the angle properly using ox bone. Dad insisted his hardheadedness was the result of absorption of products of the ox bone.

Doctor Snyder, my father, and several other Wofford professors were returning to Spartanburg by train one evening. A lone occupant at the other end of the car began cursing in a loud voice. Doctor Snyder rebuked him and insisted that he stop cursing. The man looked around and said, "Ain't no ladies here." "No," was the squelching reply, "But there are gentlemen present." The cursing stopped. After having spent a year in the Tennessee State Penitentiary (as a physician), I learned that "cussing" is an art form, and the ordinary citizen should never try it. I have heard true artists in this area of expertise use profanity for minutes on end and never repeat themselves. Do leave it to the professionals, boys. The amateur leaves only a ugly smear on the canvas of memory.

Of all the drivers of automobiles on the Campus, Professor A. Mason Dupre was the worst. On one occasion, the city fathers decided to make East Main, in the center of town, one way. (Not the first time, nor yet the last, this decision has been taken.) They installed a little box in which the policeman stood with a hand signal above indicating Stop and Go. Professor Dupre came down North Church Street and started to turn left as had been his wont. The policeman stepped out of his box, and said, "Professor, you can't turn left here!" "Young man, if you will kindly get out of my way, you will find that, with or without your help, I can turn left here." My bet is that he was allowed to turn whichever way he chose.

The Dean always inspired dread to any wrongdoer. I well recall that some miscreant had chalked over his office door in the dead of night, "Abandon all hope, Ye who enter here." The shakily written admonition stayed in place for several weeks. Years later I told his daughter, "Betty, your daddy's been dead for fifteen years and I'm still scared of him."

One of the attractions of East Main Street deserves mention. Since the railroad crossed the center of town, anything left in the middle attracted attention. A large whale had been found stranded at Charleston, so local entrepreneurs loaded it onto a railroad flat car, injected it with deodorant preservative, and trundled it through the state charging admission for it's viewing. It was placed on a spur track which ended at Spartanburg's East Main Street. Its advent had been properly advertised in the paper, but by the time it arrived in the upstate, advertising was no longer necessary. The deodorant and preservative had long since evaporated, and its presence made itself known hundreds of feet from the source. However, my curiosity overcame my nasal objections. I bravely paid my nickel to see the town's main attraction. I do recall marveling at the size of the beast. However, I did not remain long in the vicinity.

I will skip over most of the interval of teen age as much as possible. The high school years were miserable as I recall. I was smaller than most of my contemporaries, wore horn rimmed glasses, and had not achieved my prowess in tennis at the time. My peers eventually became Army Chief of Staff, General Childs Westmoreland; General of the Air Force Moore; Ambassador to Saudi Arabia, Ben Hill Brown; Father Jack Joyce, Chancellor of Notre Dame University; Carlos Moseley, President of the New York Philharmonic Orchestra; Dr. Lane Williams, head of the Department of Anatomy at Northwestern University; Dr. Moffit Cecil, head of the English Department, Texas Christian University; Judge Bruce Littlejohn, Chief Justice of the South Carolina Supreme Court, and on and on. No wonder I have an inferiority complex.

I do recall my two Latin teachers. Mrs. Evans was the wife of the superintendent of the High School. When Doctor Evans entered the front door, some mysterious message was sent throughout the building, and the whole student body became absolutely quiet until he left. Students had been known to cross the street when he approached, in order not to confront him. Mrs. Evans and Mrs Pettis, wife of the Wofford Professor, held the same aura. I do recall getting a laugh from the whole school when I offered my beat-up football helmet, (painted gold) as tribute on the funeral pyre in the Latin reenactment of death in Rome.

We had only a thirty minute recess during school but got out at two o'clock. By that time the average teenager was ravenously hungry. To get home to Wofford campus, we had to pass by the Beckers Bakery on North Dean Street. This was exquisite torture to the teen age GI tract. We hurried down North Dean Street through the underpass, and up through the woods to our house. We did not cross the railroad tracks, though we were not afraid of the trains. This red clay ditch now leads the trains under the town, but in my day it was full of tramps who lived in the numerous caves dug in the clay. (Mother could always tell when we had been playing in the "Big Ditch" because of the telltale red mud on our clothes.)

The Boy Scouts played an important role during my teenage years. Central Methodist Church had an early troop. We met in the Upper Room above the Main entrance. When the scout master was late, we would occasionally open the window which faced onto the slate roof and climb to the top of the present sanctuary. When the church needed a new slate roof after seventy-five years, it was probably our climbing that had worn down the shingles. Our scout master was one of the local dentists, and was very understanding. He knew the venturesome spirits he was supervising, and frequently took us on long hikes. On one he had obtained the train schedule for the Converse railroad trestle across the Pacolet River and took our troop over during a safe period. (Several high school students have been killed on the bridge since then, one recently.)

Our trips often took us along the rivers of this area, and I remember seeing the fecal pods floating down Lawson Fork and Pacolet Rivers when I was a teenager. It has only been in the last three years that the towns along Lawson Fork stopped dumping raw sewage into the stream. Despite the recommendation of the Spartanburg County Medical Society made just after the Civil War insisting that Spartanburg "clean up Lawson Fork."

The Depression has had a lifelong influence over both Gale's and my attitudes. While we had adequate housing and food during our childhoods, there were few luxuries. One of the most noticeable habits was the requirement to repair anything that could be fixed. If this was not possible we usually made do without it. Very little was to be thrown away. Even today this is our habit, much to the dismay of our children and grandchildren.

One time during the depth of the Depression. I thought that I had found the answer to the family problems. In a shallow pool on the edge of the campus, I discovered an oily mess, ringed by black rocks looking like coal. I could hardly wait to get my father down to look at the treasure trove, for the discovery of oil or coal would solve the problem of not enough money. Dad took one look and destroyed my dreams. It seems that this was the settling basin for the chemistry laboratories in the Cleveland Science Hall on the rise a short distance above us. I can just see the antipollution authorities of today having a screaming fit over that outdoor pool.

During the Depression there were few ways for children and teenagers to make money. One of those was offered by Mrs. Trawick who lived two doors down. She had a field of large daffodils which were among the earliest of spring flowers. The faculty children picked the blossoms, always adding one flower to the bunch of twenty-five, and we were allowed to sell them on the streets of downtown

Spartanburg. The optimal spot was in front of the Aug. W. Smith department store. (I assumed that this was because the few ladies in town, whose husbands had money, spent it in the luxury department store.) Of the twenty-five cents charged for a bunch of the flowers, Mrs. Trawick kept fifteen cents, and we acquired ten for each bundle sold.

One of the most miserable days that occurred began most delightfully. Every summer, Dad went to Columbia University in New York to complete his Master's Degree. (Incidentally he stood second in his class when he received the coveted sheepskin.) One summer we were able to swap houses for the time with a family from Bergenfield, N. J. Thus every Saturday, Dad took us to visit the sights of this marvellous melting pot. We visited the tourist spots: the Statue of Liberty, the Aquarium, Wall Street, 5th Avenue where we first experienced automated vending machines. This was sheer magic to a child from the country town of Spartanburg. Of course the magnificent museums were a must. (The musical attractions were not visited for two reasons. First, they were very expensive, and second, most of them came through Spartanburg on their way to Atlanta: Damrosch, Paderewski, Madam Schumann-Heinck, and numerous others graced the stage of Twichell Auditorium. We usually went to whatever concert was available. Our "gang" was usually among the five hundred on the stage in the Childrens Chorus. We sat and stood on tiers of planks behind the orchestra. We were taught the choruses of all the great music, and always we marvelled that the orchestra already knew the musical score which Mrs. Blackwell had labored long and hard all winter and spring to enable us to produce such exquisite notes. (We were also instructed that we were to help the child on our right if he or she got sick, which frequently happened for the auditorium was not air conditioned, and four or five hundred closely packed children always produced a faint or two for excitement.) After a Saturday spent with such delights, we were returning to New Jersey. We crossed by ferry and everybody was admiring the beautiful lighted ads on the New York side across the water, that is everybody but me. I saw only blurred colors from the distance.

As soon as we returned to Spartanburg, Dad made an appointment for me to go to the eye doctor. Since none of us had had any experience with this problem before and since I was "a big boy now," I was sent to the doctor's office alone. His office was in the new Montgomery building. After spending the morning in his office, and after having my eyes thoroughly dilated, I was given a prescription for glasses and sent home. My first intimation that I was in trouble came as I left the dark confines of the building to enter the bright sunshine. Dark glasses hadn't been invented. By closing one eye with a fist, and squinting fiercely with the other I made it back home, walking as rapidly as I dared. It was a miserable experience, and ever since when faced with a similar need, I've made sure dark glasses are available for the return journey. Being minus 400 on the eye chart is no fun.

As a teenager, it became easier to make a little money. In the summer, I was allowed to open up a drink and sandwich stand on the first floor of Old Main's right hand tower. The most frequent customers were teachers who flocked to Wofford during the summer months to work on their masters degrees. At five cents for a Coca-Cola, or Orange Crush, and the same for a sandwich, we did not retire wealthy at the end of the semester. The main drawback to the job was the requirement to ring the big bell in the other tower for the change of classes. This was an exciting chore at first, but day after day, exactly on every hour, even the most exciting of jobs becomes boring.

When the age was acquired that the business men could hire a schoolboy (sixteen, I think), my first real working job involved helping take inventory in Mr. Wallace DuPre's Automobile Parts store. This was in the old building built where the second Courthouse had stood on Morgan Square. My first job was counting the nuts and bolts in the innumerable bins up and down the length of the store. Later I progressed to an internship as salesman, really a "gofer," but I did learn the importance that one or two thousandth of an inch can make in fitting the piston with oil rings. Mr. "Mac" was supremely patient in teaching me some of the intricacies of his machine shop.

I learned about some of the realities of life also. One of my teenage acquaintances working full time in the store (he had dropped out of high school) was addicted to riding a motorcycle. Unfortunately he came over a hill on the bike in the middle of the road, at the same time a truck came over the hill from the opposite direction. It was my first encounter with my soon-to-be adversary,

Death.

One of the excitements of the age was the flight of Charles Lindbergh to Paris, in 1929. Since South Carolina had only one airport and this one was at Spartanburg, naturally he came here. The whole town turned out. For the parade I stood on the corner of what was Kennedy Place. (It was named for Dr. Lionel Chalmers Kennedy, instigator and first president of the Spartanburg County Medical Society in1868. His office was across the street on Morgan Square and became the Kenndy Free Library- the first free library in the state. It became the Carnegie Kennedy Library on Magnolia Street, the space now occupied by the County Courthouse.) I waited for and watched the open touring car pass by with my hero. Afterward Dad took us way out of town to see the famous airplane. Like most of my buddies I made a beautiful replica of the Spirt of St. Louis. I wonder whathever happened to that magnificent model?

Another childhood miracle occurred during this time. We all gathered at the new Montgomery Building to watch the Human Fly climb the ten stories. We gasped and marvelled that the daring young man did indeed skillfully climb the face of the tallest building in town. No attempt was made to stop such deeds in those days.

My father seldom lost his temper, but during the Depression he was furiously angry with the bankers of this area. He felt that they had no right to take his hard earned money from the bank deposits and gamble it on stocks. Of course, he was right, but had he been on the jury of any trial, he would have insisted on executing the criminals, instead of giving them a few years in prison.

Furthermore, he was not at all happy with the policies of Franklin Delano Roosevelt. He felt all his life that the give away programs of the Depression era would eventuate in the destruction of the United States. "When the history of the United States is eventually written, President Franklin D. Roosevelt will have done more to cause its downfall than any other one man" was his lifelong conviction. He felt, and many economists agree that the U.S. was coming out of the Depression before the "alphabet soup" of the Thirties was ever put into action. He felt that the ERA, WPA, and other "make work" endeavors were all politically inspired. Who can say he was not completely right in his statements?

I must include a Turkey episode. One Thanksgiving week, while we were living in what is now the President's Home (built by Dr. Waller) one of Mother's friends gave her a turkey. She frequently received such gifts for she had been the organist for the Second Presbyterian Church, the Bethel Methodist Church, and finally the home church, Central Methodist Church. She was also the Music Director for Floyd's Mortuary which led to her playing for funerals throughout the county This in turn led to her making many friends throughout rural Spartanburg County. (At the time of her death, age 84, she was trying to get into the Guiness Book of Records by reason of having played for 17,000 funerals.)

Occasionally during my medical practice, I received a note saying that if I would call at Such and Such Grocery Store and present the note I could pick up a ten or twelve pound broad-breasted turkey ready for the oven. No such convenience existed in the old days. When someone gave you a turkey, you got the whole bird, complete with the feathers, the innards, and the gobble. In other words a live Thanksgiving dinner on foot. On the morning in question, Mother left in a rush, saying almost as an after thought, that since I was home, and was going to be "Doctor," it would be up to me to kill the turkey for Thanksgiving. Grace, the cook had said she would prepare and cook the treat but "no way" was she going to separate that bird's head from its body. There was no recourse, Dad being at work and Mother having flown the coop. I was appointed to "do the Honors."

This Thanksgiving "piece d'occasion" was no little scrawny Turkey. He was a magnificent beast weighing at least twenty pounds, and standing tall with a loud "mate calling" gobble. His presence amid the Thanksgiving bounty was sure to be the sinecure of all eyes and appetites, and I had been given the opportunity of bringing it all to fruition. His feet were tied with a strong string, so without further ado, I fastened him upside down on the clothesline, which ran in front of the scuppernong arbor. Now clotheslines of this era were sturdy affairs. There was none of this effete act of throwing the wet clothes into a dryer as we do today. No Sir! The clothes were carefully hung on the line after being washed in Octagon soap. This combination of Octagon soap, (the strongest soap ever made! I can testify by virtue of the fact I had my mouth washed out with the soap for using a word I should never have included in my vocabulary), the blessed sun and the freshest breezes the soot laden air allowed through, completely sterilized the dirtiest clothes. Notice I said sterilized not cleaned. I had anticipated some trouble, and had sharpened the kitchen knife to razor edge keenness.

Thus armed, I grasped the victim by the neck. For those of you who have not had the dubious pleasure of closing your fist around a turkey neck, the skin of the bird is very loose and floppy. I made a rapid sweep around the neck, thinking I had severed the head from the body. My total inexperience in such deeds came through. The skin parted, pulling the floppy skin over the head, and blinding the poor bird by pulling the hood of skin over his eyes. But worse still, the ties of the feet came loose, the usually sturdy clothesline collapsed and the wily bird took off running in all directions. The fact he was blinded enabled me to corner him against the garage, tie him up again, and this time, in approved King Henry the Eighth fashion, his head was removed using an axe. This was and will remain my only experience in preparing poultry for any meal, and I am still not too keen on visiting the backyard of President Lesesne's house for the annual church picnic.

It was a different time. In so far as I can recall, none of the campus dwellings had a key. I never used a key to get in any of the three homes we occupied. One night the train from Knoxville, Tennessee, arrived at three a.m. I walked to the campus from the station, entered the front door of "Clink's" old home where we then lived, went upstairs to bed and awoke the next morning to greet the family. Incidentally, "Clink's" house is 100 years old this year.

GROUP FROM THE "SONGBIRDS OF THE SOUTH"—*The Terrier Crooners, who break hearts year after year with the same new songs.*

Back row, left to right: Prima Donna Susan Broome; Baithoven Shoebert Herbert, *Celebrated Embarrass-Tone;* Redvitsky Philipps, Handle Carooso Bennett.

Front row, left to right: Enrixo Corsaircow Copeland, Chickcowsky Storkowisky Latimer, Mozart Monroe. Director: Fido Barkinnoff.

The Glee Club

PART TWO

WOFFORD COLLEGE STUDENT, VANDERBILT MEDICAL SCHOOL, AND OFF TO PRISON

The four years at Wofford as a student were a delight. Almost all the Professors were my friends, and the studies were well adapted to my interests, so much so that I took a double major. I elected to take the AB Degree because it carried more prestige than a science degree at the time. I took several courses just for fun, such as Surveying (all really educated persons should be qualified in the art), and Electricity, which was just coming into its own.

Most homes of my childhood had a central cord hanging from the ceiling. I cannot recall a baseboard plug in any of the houses we occupied on the campus.

Early in my childhood years we had an ice box, with a hole in the floor and a rubber drain letting the melted ice water flow under the house. One of our joys was to follow the ice truck, since we were allowed to pick up the chips of ice which splintered off the big blocks from which the fifty pound household block was dissected by the expert burly black driver of the truck. He never seemed to mind our presence. Germs apparently did not exist in ice chips, for nobody mentioned them. When the ice box eventually was replaced by a GE Monitor which had a large round coil on top, we missed the iceman.

Tennis came into its own during the Thirties. For years Spartanburg was the center of tennis for the South. The mid Dixie Tennis Tournament attracted such notables as Bill Tilden, Bitsy Grant, etc. and was a grand occasion. The courts were top notch and were clay base. (I was trounced thoroughly by Bitsy Grant in the first round, but so was Bill Kennedy, the town champion for years.) My father had been the tennis team captain at Wofford, as proclaimed in the *Aurora*, the first Annual put out at Wofford. This was printed for the graduating class in 1904. Naturally, we three brothers had to do the same. In turn, I was the captain, followed by my brother Thad, (he and his friend, Francis Garrett, to become Admiral Garrett, Chief of the Chaplain's Corps in the Navy, were state champions in doubles), and lastly by my brother Elmore. One of my fondest memories was hearing the old college bell being rung as we returned to Spartanburg after beating the University of South Carolina that year in tennis. This, I think, was the only time a victory at tennis was celebrated by the ringing of the bell.

ROTC (Run Over To Converse) was almost a requirement in those days, and the numbers allowed three full companies to drill. Then came the required six weeks in camp. It was my lot to be sent to Edgewood Arsenal in New Jersey for training in Chemical Warfare. It was a delightful summer with my contemporaries from all over the US. We were equipped and clothed with the same uniforms and weapons as World War I (1918). The old flannel puttees were put on carefully before a hike, but in a short while they began to unravel. In looking up the column, one could see the marchers stuffing the three inch rollers into their pockets as the offending items became loosened.

The map reading and the firing of the mortars were fun. I became proficient enough in mortar fire to acquire an Expert Medal in that art, as well as becoming a Sharpshooter in the use of the old army .45 caliber pistol.

The Edgewood Arsenal Tennis Tournament enabled me to win a nice silver trophy over my six foot four opponent, Dog Dick Chamberlain, another college senior from the Mid-west.

I must mention the most terrifying experience of my entire life.

Dancing was of course forbidden on Wofford Campus, as was card playing, drinking, movie going, cussing, holding hands with female girls, and many other foolish negatives long since forgotten in today's overly undisciplined era. With the aid of a lovely lady, a little older than I, the handicap of never having been educated in the graces of Terpsicore, the Muse of Dance, was overcome to a slight degree: enough to participate in the Senior Ball. All that was left was to brave Dad in asking for the keys for the car for the purpose of taking a date to the said dance. My first request was a flat "no." I got dressed in my tux, and crept hesitantly down the stairs. On the newel post at the bottom, prominently displayed, were the car keys. The matter was never discussed, but I still say that coming down those stairs was the bravest thing I ever did.

The tuxedo mentioned was a luxury needed to participate in the Glee Club. The Wofford Glee Club was a delightful addendum to the campus life, and was directed by Professor Wilson Price. In the first place, we were exposed to the best of great music, in that we learned all the exhilarating choral music, such as *The Soldier's Chorus* from Faust, The Chorus from *Fidelio*, and so on, as well as the music from many of the musicals. Whcn on thc road, we were greeted as Wofford Men through out South Carolina. We were welcomed into private homes for the night, were fed sumptuously, given a party after every concert, furnished with a lovely date for the evening, and on several occasions, when we eventually retired, the bed was turned down in a neat triangle below the pillow and a delicious chocolate was left by the bedside. I've not been treated so royally since.

The Preston and Calhoun Literary Societies have almost ceased to exist at Wofford. This is truly a great loss to the students. All the students in our day were required to belong to one or the other of these great leveling societies, and each member was required to participate in a debate and to declaim once each semester. It is quite humbling to organize thoughts and vocalize them, perhaps not eloquently, but believably. It is even worse to memorize a long poem for delivery and forget the whole recital in a moment of public stage fright. The audience of one's peers eager and ready to criticize keeps a razor edge on a public appearance.

Having a passel of one's ancestors attend Wofford also tends to keep any student on his toes academically. I was frequently reminded that one of the earlier Herberts was disconsolate because he had acquired one B during his four years stay at Wofford. At last count some thirty-two men of the family had passed through the hallowed halls of Wofford. At any rate, suffice it to say that the writer was chosen to represent Wofford the senior year as candidate for the Rhodes Scholarship. This was not a success, but a national scholarship to McGill University as a result of writing a poem in French helped to compensate. I have been humbled by two compliments in past years. One I cherish particularly and only learned recently. My father told my "Infallible Critic" and lifetime companion, Gale, that he felt I was one of the few scholars he had ever taught. The other was having Wofford bestow the Phi Beta Kappa label upon me. These two have given me a lift on many an occasion.

I do not recall when it dawned on me that medicine was to be my life's work. However, it was during the high school years. (Recent studies indicate that most physicians do select medicine during the teen age.) Eventually the time came to apply to a medical school. In drawing up the application, it was necessary to supply names for future reference. What better name than Doctor Snyder, President of Wofford? He was nationally known for his scholarship and eloquence. I appeared at his office as per appointment. He was his usual gracious self and asked my problem. "Doctor Snyder, I am applying to medical school and would appreciate it if you would allow me to use your name as a reference." was my request. "And what schools have you selected?" was his response. He listened patiently as I explained that getting into medical school was difficult at that time. Only one student out of every nine applicants was selected in 1935. "I am applying to South Carolina, Emory, and perhaps Georgia." I said. Doctor Snyder straightened up and in a firm voice said "No! I will allow you to apply only to Vanderbilt." At the time, I did not know that Doctor Snyder had close ties to Vanderbilt, and that its Chancellor had taught at Wofford before his move to become the Nashveille Univeristy Chancellor. In addition, medical listings of the schools of choice indicated the rankings as Harvard, Hopkins and then Vanderbilt. Also, the school was known to be expensive even then. (Remember, this was in the aftermath of the Depression). One did not argue with one's professors in those days, so I said, "Yes, Sir," and left the office. That is how

Nashville became the next step in my life.

The exalted position of possessing a Wofford Degree today is the result of the present Wofford Faculty standing on the shoulders of the Giants of the Past. What did these teachers teach us who can no longer balance an equation, use a logarithmic table, speak French to a Frenchman or German to a German, even speak "Good English" to an Englishman? What is this arresting quality which these men endeavored to pass on - this quality which enables Wofford to survive - nay! excel in this day of mass "education" and state supported schools. It seems to me that this intangible which these teachers possessed and endeavored to pass on to us was a devotion, almost amounting to a passion for intellectual and spiritual honesty. That quality is embodied by the expression "He that sweareth to his own hurt and changeth not" or perhaps "No man, having put his hand to the plow, and looking back is fit for the Kingdom of God." (Luke IX). Each possessed in his own life that "way of life" too little regarded today. Honor! "Intaminatis Fulget Honoribus."

"We write this in order that our joy may be complete." (First John 1:4)

THE VANDERBILT YEARS

A little learning is a dangerous thing;
Drink deep or taste not the Pierian Spring:
There shallow draughts intoxicate the brain,
And drinking largely sobers us again."

– Alexander Pope
Essay on Criticism, Part I Line 15

The transition from the Wofford years to the studies at Vanderbilt was not as memorable as might be imagined, principally because everything the first year becomes a hazy recollection of intense study with no letup. I vaguely recall riding the train to Knoxville, and then boarding the unbelievable Tennessee Central Railroad to Nashville. This little soft coal burning mountain railroad hugged the mountainous spine the entire route, and even in the late thirties it was decrepit. The cars had wicker seats whose back could be swung back and forth depending upon which direction the train was headed. The heat for each car was furnished by a small upright stove in one corner at one end, and was entirely inadequate for a trip through the Tennessee mountains, especially when we came home at Christmas. The trip was particularly long because the little train stopped for any reason, even to pick up the big milk pails set out along the right of way by the local farmers, the milk pail to be delivered to the next big city, such as Cookville, population 1,000. After a long night's swaying around the curves of the ridge lines of the Tennessee mountains, the eager sooty medical students were deposited at the station, which was equally sooty, in the bowels of the city of Nashville, which was just as sooty. To those of you who have visited Nashville only in the past few years, you have missed a treat. The city lies in a bowl in the mountains, with the Cumberland River running through the middle. The principal source of energy, as well as heating, was the soft bituminous, sulphur-laden coal mined locally. In fact, the local baseball field was named Sulphur Dell. The high humidity generated by the river, the fog, and the ever present layer of coal soot made the inhabitants easy to identify, especially at the end of a day. Many times I went into the city with a clean white shirt on and returned in the evening with a filthy black collar. Those Nashvillians who came to autopsy were easy to spot, for their lungs were black with coal particles. In those days the hospitals were required to maintain a twenty-four percent rate of autopsies or lose their accreditation. A request for autopsy on every patient who died in the hospital had to be presented to the family and the fact noted on the chart. This was frequently left to the intern or medical student. It was particularly difficult to explain the need for such a procedure to the mountain folk who could barely make a mark on the permit.

We: Tom Diseker, George Price, Bill Scott, and I (all from

Spartanburg), were glad to arrive at our new home on Garland Avenue. The domicile was irreverently designated as "Pauper's Paradise," and quite aptly named too, for after the depression, we were all on a tight budget. Mrs. Compton ran a boarding house just for the "medical" students. She rented a room for two and furnished the evening meal for only twenty dollars a month, in 1936. I'm sure she made little, for we used the evening meal to cover the deficiencies of the rest of our diet. How could we put away that much food at the evening meal, and still stay awake to study until twelve or one o'clock every evening? I was a lucky first year student in that I ended up with a choice roommate. Allen Bass, a Canadian, was a graduate student in pharmacology with his Master's Degree, and teaching in that department, until he decided to pursue his MD. No one could have been a better roommate. He and I usually studied with a two pound bag of white grapes nearby, and when the grapes were gone it was a signal to go to bed. He also had to keep a tight rein on his budget, as did I. Breakfast consisted of cereal and milk in the room, while lunch was a "catch as catch can" around the noon hour. I ran across a notebook lately which had the expenditures down to the then three cents stamps for mail home. The total for the first year at VU Medical School was $743.40, and this included $125 for a required microscope. Mine was second hand. (After four years I sold it for the same price in order to buy Gale an engagement ring.)

We first year men were given a table at dinner over by the window which was usually kept open at the request of the other diners. (This may require a little explanation.) The study most feared was Anatomy. This was "make or break," for if you flunked Anatomy, you were out of school, no if's, and's, or but's. And the teachers had the ability, indeed the duty, to flunk any student who they decided would not make a "good physician." One of our most intelligent students, Lane Williams, was flunked out on this basis, and remember that this was during the era when for nine applicants to medical school, only one was accepted. (Now the requirements are so lenient that if a student applies, he or she is accepted.) Lane refusing to be intimidated by the edict, went on to Northwestern, got his Ph.D. in Anatomy and eventually became head of the Anatomy Department there.

The introduction to Anatomy is always traumatic to a bunch of carefully raised young people. Most of us had never seen a corpse before, much less having to handle one, clean off the preserving grease, and absorb the all pervading odor of formaldehyde plus the unique odor of preserved flesh. No matter how carefully one washed, cleaned under the finger nails, changed clothes and did all the other recommended deodorant rituals, the pungent odor prevailed. No wonder we had to eat at a separate table. I recall going to the neighborhood movie palaces on one of my rare indulgences. D.B. York and I sat down in the middle. In a short while, the people next to me moved, the children on the left moved, those in the back did so also, and finally we ended up as an isolated island in the dark of the theater.

Another of the inconveniences to which we never became completely adapted was the task of ridding the cadavery of all the huge roaches which accumulated after we covered "Uncle Joe." (He did resemble Stalin to some degree.) Not before or since have I seen roaches the size of those we encountered living with and on the bodies. We spent every spare moment in the Anatomy Laboratory. So much so that the authorities feared for our health and closed the laboratory on Saturday and Sunday.

One of the difficulties of Medical School was the introduction of all of us to a new language, that of medical terms. These were nearly all from a Latin base, and had to be looked up in *Dorland's Dictionary*. This was one of many occasions that previous exposure to Latin was a blessing. Of all the subjects which I have studied, I think that the language of the Romans has given me the most pleasure and been the most useful. I only wish I had been exposed to Greek also, but alas, Dad was the only one teaching the language at Wofford, and I was determined not to take any of his classes. Stupid me!

Eventually we found the delights of Percy Warner Park. This is and was a large park outside of Nashville at the end of the street car line, well kept in it's natural setting, and it only cost a dime to ride the trolley to the park's entrance. The last trolley came back to town at eleven o'clock. This granted us a whole day of traipsing in the woods, cooking outdoors, and completely breaking our routine. The butcher and the grocer were very kind to us. We could buy a nice steak, some buns, pickles, cookies, and a large can of grapefruit juice for under two dollars. There were shelters for cooking out, and retreats in case of a shower, but even the weather was very accommodating during our precious times off. I only remember one rainy occasion in the four years. This idyllic existence continued for the five or six first year students until we found that it was not an idyllic Eden without the presence of Eve. However this is jumping ahead of our story.

There were two females in our class of fifty. (Vanderbilt had the policy of taking in fifty students, losing one or two the first year, and acquiring one or two from other sources to fill the original fifty for graduation.) Miss Henrietta Freund was a delightful person and completely naive. She was the kind of person who can be sent repeatedly for the "left handed monkey wrench" or some similar medical requirement needed right now! Miss Ruth Craft was much more down to earth. Having acquired her Ph.D. in Chemistry before diverting her talents to study medicine, she was an assistant in the Chemistry laboratory. Since the feminist movement was way in the future, and Women's Liberation had not been invented yet, I'm sure these two bore more than their share of teasing and jiving than need be. They both took it in good grace however, and the class pulled together in every way we could. All of us were overwhelmed with the amount of required work.

The first time we attended an autopsy was memorable. We assembled in the stark gray room which was empty except for a bare iron stand accommodating some dozen students, and overlooking the center table which had a canvas covered shape on top. This was soon revealed by the attendant pulling the canvas off, and proved to be a beautiful young sixteen year old girl showing a most unusual yellow color. She apparently died of severe liver disease. Since this was the first time I had ever been exposed to a completely nude female and since this was the first autopsy I had ever attended, when the initial incision was made, I fainted and fell off the iron stand. Fortunately no injury was sustained, and after a few moments I resumed my original stance. Such was my initiation to the science of autopsy, and little did I know that I would do many over the years.

Organic Chemistry was and is tough. I don't know how any student without intimate knowledge of the use of logarithms could ever hope to pass, but some did. Thanks to Professor Rick Patterson at Wofford, I stood well in the class. Also, Ruth Craft's being in the class as instructor did not harm my status. Many a long hour was spent laboring over the test tubes and Bunsen burners.

I had taken my tennis racquet to medical school with me, but I don't think I ever got it out of the case that first year. I do recall stopping by the lovely field of iris which was behind the hospital and talking to the old gentleman tending them. Everyone knew the old Chancellor Kirkland who had retired. I mentioned I was from Spartanburg and had met Mrs. Blake, who was a prime iris fancier and hybridizer, as was the Chancellor. I can't recall ever having heard that he taught at Wofford before coming to Vandy. The tennis courts were just beyond the iris gardens.

This was the year that Doctor Goodpasture perfected the technique of growing viruses on living egg embryos. This meant that there was no longer need to grow the virus on the brain of a living monkey (a very expensive method of propagation). Instead of taking out a patent and selling the rights to the drug companies, Doctor Goodpasture made the use available to all scientific endeavors with no charge. This was one of the most generous altruistic gifts ever made to mankind, and one of the most important to the development of vaccines. Doctor Goodpasture and his cohorts have never been given proper recognition for their achievement.

One of the weekly events was the Grand Rounds attended by the entire student body. An interesting case was selected to be presented by the staff. This was always held in the main amphitheater, one of those large rooms sloping down to the center and arranged in a semicircle, obviously copied after the operating theaters of the Middle Ages. The history, physical findings, and lab work were presented by the student assigned to the case. He knew that he would be quizzed extensively by the Professor in charge, so he studied like mad the night before. The presiding professor was usually head of the department, be it Medicine, Surgery, Radiology or whatever entity most likely to prove interesting in the particular case presented that day. The teacher usually addressed his questions to the student giving the presentation.

On the morning in question, the patient and the information was presented very ably by the student, Mr. Carson, a fourth year man. The diagnosis was that of a bleeding ulcer, all very simple and clear. Doctor Barney Brooks was the professor in charge. Some description of Doctor Brooks is in order. He was Chief of Surgery. He was a Texan, not very tall, with a high voice which struck terror in all students, and admittedly he was the best teacher that I have ever encountered. His surgery residents were spread over the United States and held in high regard. Doctor Alfred Blalock became Chief of Surgery at Johns Hopkins and started the impetus on heart surgery by developing the technique for the relief of the "Blue Baby" syndrome, along with Doctor Taussig, the pediatrician. When "Barney" Brooks made rounds on the wards, following him first came the Head Nurse, next the Chief Resident, the other residents, the interns, the ward clerks, the janitors, and finally, way in the back were the medical students. This should be sufficient to explain the status of the medical student. The Armed Services don't even begin to approach the discipline and hierarchy of the medical school.

This day, Doctor Brooks was at his fiercest. He followed no rules. After the student made his presentation, "Barney" began his questions. Only this time, he did not question the student assigned to the case. He turned to the front row of students, which was not in the program at all. Pointing his finger, and demanding in his high, whining voice, he chose a totally unsuspecting and totally unprepared Mr. Brown. "Mr. Brown, this man has an ulcer obviously. What would you feed him?"

Mr. Brown gulped, and whispered loudly, "Sweet potatoes, Doctor Brooks." Doctor Brooks bellowed, "My God, sweet potatoes!" Whereupon Mr. Brown fainted, slid out of the front row seat onto the cold concrete floor of the amphitheater. Some of the students in the front row started to aid their stricken fellow classman, but "Barney" waved them back. He continued his lecture as if nothing had happened. In a very short while Mr. Brown began to recover, raised his head, and "Barney" said again, " Mr. Brown, about these sweet potatoes!" Mr. Brown laid his head back down, closed his eyes, and the lecture proceeded. As I said, Doctor Brooks was one of the best teachers I ever encountered.

Mention of the amphitheater reminds me of one of the noticeable changes in customs in the past fifty years. It frequently happened that the windows of the theater had to be opened to the outside (remember that this was before the days of air conditioning) in order to allow the tobacco smoke to dissipate. The air would become so thick due to the doctor's smoking during a meeting that the light from the slide projector could not penetrate to the screen . Smoking is no longer allowed during public gatherings, and most physicians no longer use tobacco.

This has been a big change.

Doctor Blalock was a rather aloof character. He took his medical duties very seriously. When he lost a patient he would frequently withdraw into seclusion until he was emotionally prepared to resume his duties. He was very meticulous in his preparations for operations, especially in the realm of asepsis. He was very strict regarding operative technique. On one occasion, an old Tennessee physician referred a patient for operation, a chest case. He was invited to attend the operation, but unfortunately he was late. The operation was well along when he appeared in the operating room. He did not bother to change clothes and proceeded to enter the operating room in his street clothes. Doctor Blalock acknowledged his presence without looking around, and having exposed the pathology, called upon the old man to look into the chest wound. When the old man leaned over, his tie fell into the opened chest. This caused Doctor Blalock a great deal of anguish, which he expressed volubly. To which the old gentleman replied, "Now, Son, don't carry on so. It was an old tie anyhow."

On one of the infrequent trips around the beautiful Tennessee countryside, we went to visit an old physician, Doctor Walker, who lived and practiced in the Mule Capitol of the Universe, Franklin, Tennessee. (Or so the big advertisements to the fall fair claimed.) Doctor Walker professed to have given the first smallpox vaccination in the state of Tennessee, and, after meeting and talking with him at length, I was inclined to believe him. He told the tale of a young lady who had been in labor with her first child for a long while, and he had a vial of a new medicine called Pituitrin (only he pronounced it "Pyetuitrin") available. Now this drug does produce violent uterine contractions and is used in minute doses, beginning with never more than one drop as the initial dose. So Doctor Walker gave the laboring patient a shot of "Pyetuitrin." "How much did you give her, Doctor Walker?" was my question. "I guv her the whole vial" was his response. She evidently delivered shortly afterward, for my next question was "Did she tear, Doctor?" "Son, she tore clean to her shoulder blades!" he answered.

Physiology was an interesting course of study, in that the laboratories occupied a large part of the discipline. (I am still amazed at how long a frog or turtle's heart will continue to beat after being removed from the animal.) These hearts would be subjected to various drugs to

WILLIAM C. HERBERT, JR., *A.B.*
SPARTANBURG, S. C.

Senior Order of Gnomes, Χ Β Φ, Β Π Θ,
I. R. C., Blue Key

"One equal temper of heroic hearts,
Made weak by time and fate, but strong in will
To strive, to seek, to find, and not to yield."

First Year: Distinction in Scholarship; Freshman Debating Team; Freshman Y. M. C. A. Cabinet; Glee Club; Monthly Orator Preston Literary Society; Private R. O. T. C.; I. R. C.

Second Year: Distinction in Scholarship; Laboratory Cabinet; Beta Pi Theta, French Fraternity; Preston Literary Society; Chi Beta Phi Scientific Fraternity; Glee Club; Corporal R. O. T. C.

Third Year: Distinction in Scholarship; Laboratory Assistant in Chemistry; Winner National Prize for French Poetry offered by Beta Pi Theta; Y. M. C. A. Cabinet; Historian Junior Class; Monthly Orator Preston Literary Society; Tennis Team; Winner Military Tennis Tournament, Ft. Hoyle, Md.; I. R. C.; Chi Beta Phi; Blue Key; Color Sergeant R. O. T. C.

Fourth Year: Distinction in Scholarship; Rhodes Scholarship Candidate from Wofford College; Editor-in-Chief Old Gold and Black; Laboratory Assistant in Chemistry; Senior Order of Gnomes; Captain Tennis Team; VicePresident South Carolina College Press Association; President, Vice-President, First Critic, Senior Monthly Orator Preston Literary Society; Cadet Captain R. O. T. C.; Y. M. C. A. Cabinet; Vice-President Chi Beta Phi; Chairman Lyceum Committee.

William C. Herbert, Jr. - 1935

Snyder Football Field - 1935.

Captain
Wofford Tennis Team '35

Brother Thad
1935 Spartanburg City Champion

Wofford Tennis Team - 1935

Field Training - Edgewood Arsenal - Chemical Warfare School

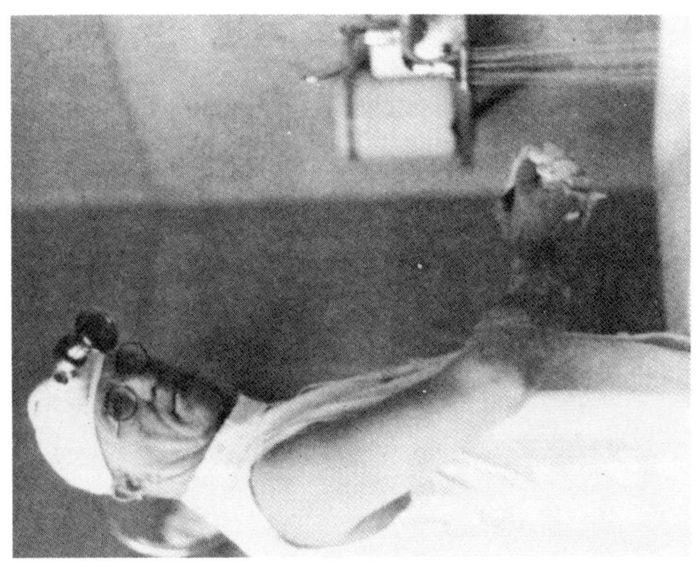

Dr. "Barney" Brooks Vanderbilt
Chief of Surgery

Chemical Warfare School
Dress & equipment from World War I.

Study Area - 1939

The author and Hudson. Pride and joy.

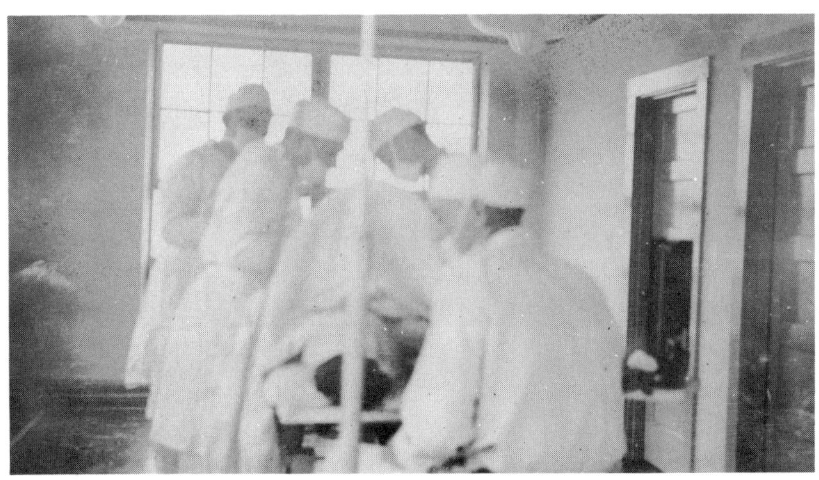

Operating Room at Tennessee State
Cordell Gode, Dr. Magruder, the author and Bill Adams - anaesthesia.

Dr. Goodpasture, Head of the Pathology Department
Discoverer of the Culture of viruses upon egg embryos.

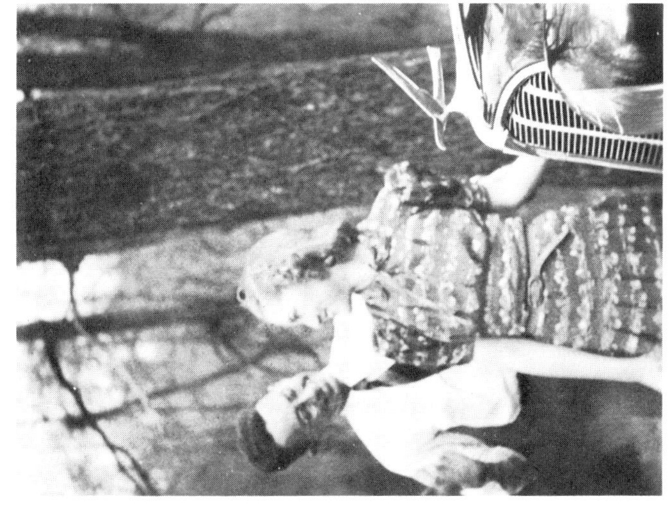

Fun days at Percy Warner Park
The author, Gale and the Hudson.

Vanderbilt Graduation - 1939
A "sho nuff Doctor" and a real "Girl!"

"The Duke," D. B. York and "Bill" Herbert on the steps of the Tennessee State Penitentiary

Professional Staff of the General Hospital, Tennessee State Prison - 1939

determine the efficacy of the medicine. I wondered at the time how long the human heart when removed by the Aztec Priests at the time of Columbus continued to beat after removal. I have never seen an account of the sacrifice which mentions any time, or whether the heart was still beating.

Pathology I passed because I had learned to draw from Doctor Cook at Converse one summer. I was able to look in the microscope and reproduce the pretty bugs and cells with the aid of sharp crayons, though I had little notion of the pathology involved or the disease process the deadly bugs induced.

The third year was the time in which we became clinical clerks, the year when we were first introduced to the patients. They were as a whole very patient and tolerant of us. Every patient, no matter what the financial status of the individual, was assigned a medical student. (The admitting physician was always very careful to be present and supervise the physical examination of a valued Nashville bigwig.) We were also required to do the laboratory work on the patients.

How well I remember the consternation I caused one day when the vaginal smear of one of the society belles turned up to have intracellular gram negative diplococci on the slide. I quickly called the resident to verify my findings, and he called the pathologist. They all agreed that this was the finding and reported that the smear indicated that the diagnosis on the social butterfly had to be Gonorrhea. How this was handled by the private physician in charge of the case I never was informed. This case should have been discussed in the open amphitheater before the student body for the edification of the medical students in how to tell a patient of the upper crust that she has a venereal disease. Strange to say the case was never mentioned again. Ahh! The perks of money and social status will always prevail, even in a communist society.

Another lady stands out in my memory. While I was a clinical clerk, a mountain lady was admitted with a diagnosis of pneumonia. During the taking of the history, the question was asked if she had ever been admitted to Vanderbilt Hospital previously. She volunteered that she had been in the hospital some twenty years before with cancer of the breast. Examination showed no evidence of the cancer at this admission. The previous record was produced, and it seemed that the woman had been seen for cancer of the breast. She had refused all treatment, except she had allowed the physician to make a biopsy, under local anaesthesia, of a lymph node from the left axilla which was filled with nodules. This biopsy confirmed the diagnosis of cancer of the left breast with extensive metastasis. She had turned up some twenty years later with no evidence of breast cancer.

This story reinforces the feeling that we don't really know very much about the mysterious disease of cancer of the breast. The death rate from this cancer is the same as it was some hundred years ago when Halsted introduced the radical operation of removing the breast and the axillary lymph nodes, though the five year survival rate has improved somewhat using mammography. The mortality line on the American Cancer chart shows an almost straight line for a hundred years.

This was the year I left the Pauper's Paradise for a place up the street. Again Tom Diseker had preceded me. I fired the furnace for two little old ladies. I lived in a (ten by twenty feet) lean-to built onto the garage. There was a wash basin on one side, an antique commode in the corner, covered when not in use by a large wooden box, a cot along one wall. It was heated by an electric heater. At best, it was Spartan but barely adequate.

At Christmas time all the Spartanburg students piled into Tom Diseker's car. (He had to have one in order to get to class from the penitentiary each morning.) This year we hit the high mountains near Knoxville about midnight in the midst of a snowstorm. We tried in vain to push the car over the last pass. Eventually we turned around and trekked back to the town in the valley, where we finally located some chains and crossed the snow-hidden pass to arrive safely in Spartanburg the next morning.

This was the year the Ear, Nose and Throat Department thought my tonsils and adenoids should come out. With children this is a minor procedure, but with adults, it is a different matter. The operation took place in the morning, but for some reason the wound continued to bleed all day. It ended up with my receiving two units of blood and the resident, who incidentally had tremendous hands, suturing the offending vessel without anaesthesia (as my dazed state recalls).

This was also the year that the medical school required that all third year students deliver five babies in the homes of the indigent women who attended the clinic. When the call was received, the two innocent medical students were given two medical bags full of linens, gloves, needles, sutures, scissors, cord ties, one can of ether, etc. Finding the patient in the wilds of Nashville was no small adventure in itself.

On our first delivery call, which was in January, two of us went into the hinterland outside Nashville to a log cabin out in the woods. The lady had already had two other children so she was a lot more educated in the procedure than we were. We spread out all our professional equipment, requested that the water be boiled (incidentally, I never did find out why all that water needed to be at hand), and settled down for a wait. We timed the contractions, listened for the fetal heart and did everything to pass the time away.

After a long long time (actually about two hours), the lady said "I think it's about ready to come." I put on my gloves, rearranged the drapes and made sure the bed and floor were covered with newspaper. My partner poked up the fire and reset the oil lamp, for it had become quite dark by then. We settled down to wait patiently. After awhile the patient announced the baby was coming. I stood where I never should have been standing for when the water broke I was drenched from head to toe with amniotic fluid. The fine boy cried lustily, the cord was tied, the baby washed, the placenta delivered without incident, everybody was happy except me. I pealed off the wet shirt and undershirt, and my partner let me have his sweater, all the time laughing excessively, I thought. One further indignity remained. Tennessee State Law required that all babies eyes had to have a 1% silver nitrate solution instilled into their eyes to prevent infection. Since I had been inundated with the same amniotic fluid, it was necessary to insert some of the same into my eyes. Nobody had ever asked any of the babies if the procedure hurt. Believe me, silver nitrate stings like Hades. My partner had to clean up the debris and drive home.

Though we may have been tempted to name some of the babies we did not follow the example of one of our classmates who was asked to suggest a name for the twins he had happily delivered. His answer, adopted by the mother was Sifilus and Gonora (Syphilis and Gonorrhea). There are some babies named Herbert, both here and in Nashville, in honor of the obstetrician.

On our final year the Duke of York, (i.e. D.B.York) and I acquired the choicest job which fourth year students could hold. I continually startle my acquaintances at times by announcing that I spent a year in the Tennessee State Penitentiary, and so I did. Doctor Frank Brown, from Goldsboro, N.C. was a recent graduate awaiting his internship, and acted as Chief for a few weeks. Doctor Tommy Frist, an outstanding Internist, and Doctor McGruder, both on the teaching staff at Vanderbilt Medical School, oversaw our work and made daily rounds at the two hospitals. Doctor Frist set up and now runs the tremendously successful private hospital chain headquartered in Nashville.

There were two hospitals within the grounds of the Tennessee State Penitentiary, a maximum security institution, and the site also of electrocutions. The hospitals were forty beds each, one a general hospital and the other was devoted to the care of the numerous tuberculosis patients. The hospitals were staffed by the inmates, and the personnel were well trained. I seldom noticed a break in technique, and any of our corrections were well received. These were the "cushy" jobs inside the penitentiary and were eagerly sought.

Our secretary at the main hospital was a highly educated man who was in prison for forging and selling Tennessee State Bonds on which he made a fortune. He once asked me for a pass to the prison farm down the road, which happened to be next to the women's prison. I, as a matter of course refused, whereupon he challenged me to write my name on a piece of paper and turn my back for a minute. When I turned around he had copied the signature on ten different pieces of paper so exactly I could not determine the original from the forgeries. Having a sense of humor, as well as accepting the situation, I wrote out his requested pass.

The technician in charge of X-ray was a very small man, but highly intelligent. He was incarcerated because he had eventually been caught at his trade as a second story jewel thief. He very thoughtfully offered to supply a two carat engagement ring when the news of Gale's and my engagement made the news. The price for such a ring was unbelievably low, and this might have raised questions as to how he had acquired such a ring. I never asked, for the inmates were never guilty of the crimes assigned to them, but always "had been convicted unjustly" of the alleged infraction of the rules of society. I cannot recall any prisoner having any guilty conscience over anything he had done.

The new Doctors were always a target for any of the prisoners, especially "sick call." Any excuse was used to get out of the cell. Therefore when new interns arrived the number in the sick line rose from an average of twenty or thirty a day to four hundred. Something had to be done. We had to be in class at eight in the morning or even earlier if surgery was scheduled, so sick call was at six o'clock. The previous fourth year students at the prison had warned us of the avalanche of patients and how to cope with it. They had worked out a very ingenious system. Every other man in the line was given one drop of a very sticky solution of Tincture of Valerium which was carefully inserted in his nose. This drug has a very foul and persistent odor which not only persists but notifies every person around to stay as far away as possible. In the close confines of a prison, this is not always practical, so the sick line dropped to an acceptable number in a very short time. The drug has no ill effects.

There were quite a number of men who had persistent Gonorrhea. (this was some five years before the advent of penicillin). and numerous cases of urethral obstruction which had to be dilated. The Duke of York and I quickly became adept at passing the very small filiform minute catheters to relieve overdistended bladders caused by the adhesions in the urethra. Even though we inserted an anaesthetic jelly before the event, the resulting dilation was remembered by the patient and the surrounding sympathetic patients. I wish that I thought the incidence of GC. has dropped because of the painful treatment, but I think the judicious use of penicillin has had a greater effect. (Incidentally, years later in St. Paul's Cathedral in London, I was determined to look up the grave of Sir Alexander Fleming, who discovered penicillin. We went down into the crypts of the building, and were quickly surrounded by the monuments of the great and near great. Nelson's tomb dominates, being closely approached by Marlborough. The Fleming plaque measuring three feet by two, lies modestly against the wall out of the way. "Sic transit gloria mundi." What a puny way to honor one of mankind's greatest benefactors!)

On Saturday night the upper cells became the Red Light District in the prison, and nobody in the administration or the guards would enter that section. Whatever happened was the business of the prisoners. At least we the physicians were not called to attend any of the prisoners during that time.

This accounted for the large number of cases of syphilis seen in the prison population. We had almost four hundred in the "shot line" on the one morning each week when the arsenical intravenous medicine, Salvarsan, was given to the bearers of the dreaded spirochete. It was supposed to have taken a full two or three minutes to administer each dose, but with the line so long, and the clock ticking for the eight o'clock class I'm sure this was not observed all the time. This time pressure made each of us very adept at hitting the vein. Only one case during the year there in the penitentiary developed a severe reaction to the medication. This evolved into a terrible dermatitis which progressed into a purulent staphylococcus infection from which the patient eventually died. We even tried the new medicine, Sulfanilamide, without avail.

On only one occasion was I ever apprehensive for my safety. Since the penitentiary was about ten miles out in the country from the medical school, a car was necessary. Dad had been able to procure the beautiful Hudson coupe which had belonged to Doctor Blake. The good Doctor had died and Mrs. Blake had the car put up on blocks and properly cared for. It was in pristine condition. One of the perks of the job was thirty gallons of gasoline a month plus having the car cared for by the inmates. They were always eager to work on the car for it had the newest gadgets, including a new electric gear shift. This was in 1939. A little lever beside the steering wheel, about two inches long, was the means of shifting gears, and this activated a clutch at the same time. Far out advanced technology for that time.

The car needed the oil drained, greasing and general care, so I brought it in. The entrance was by a large gate into a cul-de-sac, where the guards inspected it after closing the gate. Then the inner gate was opened, and the car allowed to go to the cavernous garage. When I came back to pick up the car, one of the prisoners called out, "Commere, Doc. I got sumpin to show you." I went back into the dirty shop in the back. When I arrived he pulled out a great big knife with a blade a foot long. My first thought was "What have I ever done to this fellow?" However, he handed me the knife which had a lovely handle made of bone alternating with pieces of toothbrush handles. The blade was glaringly sharp. He had made the "shiv" by grinding down an old file. How many hours it must have taken! I finally convinced the man that I really had no use for such a lovely knife. I settled by buying a ring made of toothbrush handles which has two small dice complete with the proper numbers. Gambling was forbidden, but with this tiny dice set a man could turn the ring inside his palm and nobody could be any the wiser. I still have the ring as a memento of an exciting encounter.

The prisoners used every device to become patients, for the boredom was less, the food better, and it was a much more congenial in the twenty bed ward than in the cell for two. One of the devices was hard to detect. A rather skinny man was readmitted for vomiting blood. X-rays did not reveal any gastric or duodenal ulcer, and we were on the verge of sending him back to his cell when he vomited a large amount of blood again. This went on for some time in spite of a high index of suspicion that he was malingering. Finally one of us caught him inserting a needle into a vein, swallowing the blood and then regurgitating it at a psychologically opportune moment.

We the fourth year students, did all the appendectomies, the hernias, sewing up the numerous lacerations, and on one occasion I was given the job of amputating an arm. Lonnie Taylor, one of the prisoners who was set for execution for a particularly brutal murder, almost made it over the high wall, but he was shot in the arm and fell. The wounded arm became infected and developed gas gangrene. This dread disease is caused by a deadly Clostridium bacterium, and at that time the only way to prevent sure death was to remove the limb. Doctor Magruder was there to supervise this one, and the left arm was removed without incident, but the patient died of the subsequent septicemia.

Since the dentist came but once a month, of necessity we had to do any emergency dental work. On one occasion, a huge, well proportioned negro man presented himself expressing severe pain in his left upper incisor, and insisted that it be pulled immediately to relieve the pain. I could find nothing to substantiate his claims, but since he kept insisting that something be done, I pulled the unoffending tooth. In a few weeks, I saw him in the yard and he flashed me a tremendous grin, which afforded a generous view of a gleaming gold tooth in front. The gold tooth was further embellished by a diamond in the center of the tooth. I still wonder if I was hoodwinked into pulling a normal tooth.

Why we, as medical students, were chosen to witness the executions in the electric chair and then certify the prisoner as being legally dead has forever eluded me. That was part of our job, and, while memorable, it was none the less, emotionally disturbing. I was required to attend three while in the penitentiary. Even after the current was turned off and everything was quiet, entering the death chamber and using the stethoscope to listen for heart sounds was not an easy task. Being in the atmosphere for a year did indeed fix my attitudes however. I am a believer that execution is a strong deterrent force. Our error is in the slowness of the enforcement. The two deterrent forces are the certainty of getting caught and the swiftness of punishment. In neither of these does our system of justice approximate the ideal. I also firmly believe that about fifteen percent of our criminals are not capable of ever conforming to the laws of society. The rate of recidivism among released criminals confirms this observation. I also feel that the same percentage, around fifteen percent, are ineducable. It is high time that society identify these individuals and rectify the situation upon which so much of the public funds are wasted. Since I am on a soapbox, let us extend this further. I do not believe that fifty percent of our population can benefit from a college education. At present too many receive a college degree, which is not the same, by any manner of means, as a college education.

The studies at Vanderbilt became less pressing as the duties of the prison intruded. We attended many Grand Rounds. Some of the most memorable were conducted by Doctor Tinsley Harrison. He was, without doubt, the nearest of all my teachers to approximate a computer (which was not invented for another forty years). His presentation of the approach to diagnosis has not been challenged. It was a work of art to see him fit the puzzle of a weird diagnosis together and come out with a correct and logical answer.

It is hard to believe in this free sexual era, but it was not until our fourth year in medical school that we were permitted to peruse the heavy tomes of Havelock Ellis, Van der Velde, Freud, and others of the intellectual greats because of their far out theories in sexual psychology. Not only that, we were permitted to read these volumes only in the Medical Library, and they had to removed by the librarian from a locked cabinet and handed to the individual reader. I would not have been surprised if she had asked for a letter of permission from my parents.

At last came the great day! Not graduation but oral "checkoffs." If we managed to pass these dread examinations we were pretty well assured that a diploma would be forthcoming. And as luck would have it, I was assigned to be interviewed by none other than Doctor Barney Brooks. It wasn't so much that I was petrified by Doctor Brooks, I was tongue-tied in his presence. To those of you uninitiated, the oral check-off consisted of an interview on anything the professor chose. His prerogative was to question the student about any subject which the student was supposed to have learned in the past four years in the medical school.

Now it was well known that Doctor Brooks loved to humiliate a student by asking intricate questions about gallbladder disease, so we studied the gallbladder and liver like mad. His only son had died tragically of osteomyelitis in the mastoid some years before, so we were prepared to answer questions relating to this disease. Came the big day. I ate breakfast very sparingly, for I did not wish any interference to be occasioned by my GI track. I was ushered into the great man's office by the secretary. He was very gracious, which was viewed as a very bad sign. When "Barney" was nice, better watch out, you could be in real trouble.

His introductory "Harrumph" as he cleared his throat sent shivers up my spine. In his querulous, high pitched voice he asked the first question, "Mr. Herbert, what do you know about reasoning by concomitant variables?" Since I knew absolutely nothing about "concomitant variables" and had never heard the words before, I managed to truthfully reply, "Nothing, Doctor Brooks." Whereupon he proceeded to give me a thirty minute lecture on the frequent and fallacious use of reasoning by such variables. While I appreciated the valuable instruction, of which I remember very little, my main prayer of thanksgiving still remains that he did not ask me any other questions. Only later did I find out that my wife-to-be and Doctor Brook's daughter were good friends, and Gale had spent many nights in "Barneys" house. Also my future mother-in-law was a Texan, as was "Barney," and they were well acquainted, since both of them hailed from the metropolis of Seymour, Texas, a town of around 2,000 persons at that time. Again the old adage, "It ain't what you know, but who you know," was at work. I have always had a firm belief in serendipity.

In the penitentiary, we had all our meals in the Tuberculosis

Hospital. I guess that the choice was limited for that was where the kitchen was. The patient's meals were sent down to the main hospital by carts. Our cook was Georgia Boy, who was incarcerated "unjustly" for carving up his wife and her boyfriend. Now Georgia Boy was a good cook as long as the food was in the Tennessee and Georgia menu. He specialized in grits, ham, red eye gravy and a few other of life's essentials, such as big biscuits and fried eggs over hard. Cholesterol had not yet been invented. Now one day I made the mistake of inviting two lovely guests to dinner. Georgia Boy was eager to prepare a feast. The Duke of York and I escorted our guests through the main gates and up to the dining room. Everything went well until Georgia Boy brought in his "Piece de Resistance," a large baked hen. I never did find out what he really did to that old bird, but when I tried to plunge the fork into the breast of the chicken, the fowl flew across the room. A second attempt was equally unsuccessful, so "Georgia Boy" took the offending meat course back into the kitchen, and finally hacked it to pieces with a meat cleaver. He was properly humiliated for the next few meals.

To get to the TB Hospital we had to go through the main prison yard, frequently during the time of the prisoner's exercise period. One of our visitors, Venable Vermont, later to be judge, notes that though the prisoners were forbidden pets, there were numerous small "deskunked" skunks wandering about the exercise yard.

The noon meal had to be eaten close to the medical school, for we were on the wards and in class most of the day, and hence a good, cheap lunch, rapidly served, was in order. We eventually discovered that the Peabody College cafeteria met the criteria, beside the added advantage of having lunch, surrounded by pretty girls. This in time eventuated in our being asked to the Peabody dances. One evening in the spring of 1938, three of us from Pauper's Paradise were invited to attend a spring dance. My civilian clothes may have been a problem, for my two suits were old ones from my Uncle Roy, a physician in Norfolk who was quite wealthy, not from doctoring, as many thought, but for having the forethought to buy five shares of original Coca-Cola stock early in the game. My mother had cut the suits down to fit me, but they still were clothes more fitting a mature man, and completely out of style.

At the dance I zeroed in on three lovelies. One of them was a sexy wench, whom I found out later was rather free with her female gifts. The second was a beautiful girl, but seemed somewhat aloof. I found out later that the feeling was widespread, that nice girls didn't go out with medical students. The third of my attention that night I would have given up on completely, for when she and a good looking boy started dancing, all the other dancers stopped to watch them. Of all my accomplishments, dancing is definitely not one of them. She and her brother, for so he turned out to be, danced like professionals. Luckily, the beautiful dancer was in charge of the arrangements for the dance and among her duties was the one of seeing that the guests had a good time. A mutual friend introduced her, but in those days, the stag line ruled. Our dances were all too short, but I did get her address and phone number.

Late one evening the next week, after I had done all the studying I had planned to do, I called her. She too had finished her studies. I hastened over. Dark clouds were lowering in the west, so I mentioned, "Lets go up Reservoir Hill, and watch the lightening." So we did. The display was well worth the walk, but we did not pay as much attention to the approaching storm as we should have. The heavy rain caught us on the way back. This wasn't too much of a catastrophe, except that Gale was wearing a brand new wool dress and it shrank so much that it never could be worn again.

The next time I called, her father made chit chat in the living room until she was ready to come down. One of his questions was "Do you play chess?" I did, whereupon he challenged me to a game, the pieces of which he just happened to have set up beside us. He was a good player, and I was determined to do my best against the father of my date. An hour later we were still playing while my date fumed in the corner beside us. I think I declared a draw before the storm hit from Gale. I found out later that this was one of her father's favorite methods of teasing her.

Eventually, the four of us, D.B. York, Alan Bass, Joe Car-michael and I, became bold enough to ask our dates to go to Percy Warner Park with us. It was too special a place to be shared with anyone who might not appreciate the beauty and silence of the hills of Tennessee. Gale proved able to hike and climb with the best of us. Eventually we settled into one of the numerous shelters and cooked our steaks, drank the can of grape juice, ate the cookies and white grapes we always brought along. The silence settled, until one of the others said, "Bill, sing for us." Gale shrank back, and mentally said "Oh, no, not that." But when the deep base notes of "On The Road To Mandalay" and "Deep River" came echoing back from the opposite hills, she decided she would not be too embarrassed after all. We all sang the old favorites, and I ended the session with the "Soldiers Chorus" and "Old Man River." She has never had any reservations about my singing since. (Our tastes in music were quite eclectic. In fact, we even went to the Old Ryman auditorium one time to hear and join in the Grand Ole Opry. Once was enough.)

In 1937, the Cumberland River overflowed its banks, and a great flood resulted. (This was before the building of the numerous dams of the Tennessee Valley Authority controlled the river.) All the physicians served double duty, and even the medical students were pressed into service. This was the first time I had left the sacred halls of Vanderbilt Hospital and was exposed to the disorder of a large public hospital. The insouciance and laissez faire attitude was a real shock. The flood victims survived the antiquated equipment and the lack of real cleanliness.

All was not study and grind as the pictures show. The parties at the medical school deserve a mention, as they have had and still have a somewhat dubious reputation, well deserved no doubt.

The first party to which I took Gale was held in the old and revered Maxwell House Ballroom in downtown Nashville. The festivities were in full swing when we arrived and a large amount of alcoholic beverages had been consumed, but decorum still prevailed. After an hour or so of dancing, we noticed a circle of dancers were collected in the center. On approaching, we saw one of my classmates sitting in the center of the ballroom floor, clad only in his shorts, a towel wrapped around his head as a turban and "playing" a clarinet. His active imagination plus numerous trips to the punch table led him to believe the snakes he was seeing could be charmed by clarinet music. They apparently did disappear, since he was unceremoniously hauled off feet first, after passing out on the dance floor.

This was the era of scavenger hunts, where the participants were given a list of numerous items to be obtained by any means fair or foul, within a limited time frame. Now it so happened that a Charleston boy was on the Surgical Service at Vanderbilt, and was a participant in the famous scavenger hunt. The following morning a disturbed Tennessee farmer appeared on the doorstep of the Frat house. On questioning he said "I want my pig." It turned out that one of the requirements for the hunt was a pig, and the farmer said that Doctor P— had promised to bring back the little porker the next morning, and "By golly, here it is almost noon and the pig still hadn't been brought back." After much searching the sleeping Doctor was discovered in his bed and beside him was the missing piglet. From then on the good doctor was known as Piggy P—. He still is called Piggy in certain circles, with appropriate "oinks" but never, never by medical students in the Medical School of South Carolina, that is if they know what's good for them.

At the time I graduated, the powers in control had decreed that in order to pursue further studies at Vanderbilt, it would be necessary for the new M.D. to have a year of rotating internship at another institution. A rotating internship meant going to some hospital which offered three months on Surgery, three on Medicine, three on Obstetrics and Gynecology, and three in the Emergency Room. Such a position was offered at the Grady Hospital in Atlanta, run under the auspices of Emory University Medical School. The old Grady was the teaching hospital of Emory. It was located in mid town Atlanta, right where the Marriot Hotel is today. Backing up to it was the old Opera House.

The medical rotating internship was a brutal one. On one count, I was responsible for a ward of men numbering fifty-two patients. (The present rules require that the intern never have more than twenty-five at one time.) With that many ill patients, the man responsible hardly knew the names, much less the illnesses and needs of the sufferer. We were housed in a small room with four physicians in the room. We slept in double decker bunk beds, when we had the luxury of sleeping. The pay was ten dollars a month, with board and uniforms furnished. The fatigue, the demands, and the overcrowding resulted in the Grady having the highest incidence of tuberculosis among it's house staff of all the hospitals in the United States.

On a lighter side, the Grady Emergency Room always provided a continuing drama, be it comedy, tragedy, excitement, fast paced life, and humdrum waiting. By law, the two officers driving the ambulance, had to be accompanied by a physician. This way we got to know Atlanta from the gutters to the wealthy, all of it from the seamy side.

On one occasion, while working in the ER one Saturday night, I was sewing up a nice negro man, who had been cut up in some fracas, and he asked me, "Would you like to come to the Opera next week?" Not having been exposed to opera previously, and being intrigued to know how the humble black man could get tickets to the event, I showed a good deal of enthusiasm about going. The Metropolitan Opera came every year to Atlanta for a week of "couth" to the benighted South, and tickets were very hard to come by. His reply was, "Well, you know the Opera House backs right up to the Grady Hospital, and every night I leaves the little door on the right side open a little bit. If you wuz to go up to the third balcony and stand beside the post there, nobody will say nuttin. But when the light goes on, you hafta go to the men's room on that floor, and stay there 'til the lights go off again."

Evidently the conspiracy had been going on for years and was accepted by the hierarchy. This year the "Met" had decided to put on Wagner's *Ring of the Nibelung*. To the uninitiated, this opera consists of five consecutive nights of some of the dullest repetitive stories and music ever put on the stage. I did not know this at the time, so I arranged with my fellow Interns to cover for me during the performances. Came the big night. I put on my freshest uniform, slipped in the back door of the opera house, and made my way to the highest tier. There I went into the men's room and waited until the music started. The first act went well with the dwarf stealing the enchanted gold from the Rhine Maidens. When the curtain descended, I retreated to the Men's room again. This time I had company, but no untoward remarks were made. After the lights came down I ventured out to take my place beside the pillar as before. This time however, somebody had left a Coca-Cola bottle beside the pillar. I kicked it. As some of you remember, the old bottles were covered with lovely ridges extending from bottom to top. Right in the middle of the opening aria, that confounded bottle rolled all the way down the cement steps, from the top to the bottom, never breaking, making "Brrrrr Bump" all the way down the seemingly interminable steps.

In spite of this embarrassing episode, I stuck out the five nights of the *Ring*, and have been addicted ever since. Another Saturday episode stands out in my memory. A young black girl lay on the examining table in the Emergency Room of the Colored Hospital when I came in. When I said "What's wrong, girl?" she snatched away the sheet to reveal the fact that she had no clothes on below the waist. However, she had numerous small cuts between her legs up to the pudenda. While I was engaged in the tedious job of suturing the cuts, I asked her (purely to make conversation), "Where are your underclothes?" Her reply was "Law, Doctor, my mammy said there warn't no use my wearing them when I left home tonight since it wuz Saddidy nite."

Undoubtedly the most embarrassing episode of my life occurred during the year at the Grady. My enduring girl came down to Atlanta to enjoy the opening of *"Gone with the Wind,"* which was complete with Clark Gable and Vivian Leigh in person to make sure that the long movie was a success. It was! But the following day was Sunday, and we had decided to go to the Methodist Church (the old church is now completely surrounded by Sex Shops in the center of Atlanta) to hear Doctor Pierce Harris. To make it to church, I had to get up early, make rounds, change to "civies," and make it to pick up Gale. I skipped breakfast. We were a little late getting to church. Now in the Methodist Church, late comers are escorted to the front seats, the back seats all being occupied by the regulars. In a little while, Doctor Harris began his sermon. His rather gravelly voice was accentuated by the gravity of his sermon. "The thundering tones of war are rolling across the continent of Europe today." was his introductory remark. My empty stomach emphasized his statement by thundering also, accentuated by the wooden pew back, which acted as a sounding board. In a few minutes, Doctor Harris again referred to the thunder of war, and again my stomach thundered forth, aided by the wooden pew back. I had no recourse but to glare at my seat mate, and move slightly away as if she were the culprit. She has eventually forgiven me but not completely.

The colored hospital at the Grady was an old building and run-down. The Labor and Delivery Suite - no, not suite but room was on the top floor. This was a large room with five delivery tables, lined side by side across most of the room and facing the large open windows, no air conditioning, of course. The laboring patient was brought from the ward and placed on one of the delivery tables as the time approached. In those days, the drug under investigation was paraldehyde, with sometimes scopolamine as an adjunct. Now the paraldehyde was mixed with an oil to prevent it's irritating effect and given rectally by enema. The scopolamine (you know it as the "truth drug") was by injection. This combination was guaranteed to produce amnesia, and most of the patients remembered nothing of their labor, their pain or the advent of the newborn. However, the medication also produced a complete loss of inhibitions. As a consequence of the labor pain and the drugs, the mother-to-be would leap from the table upon perceiving a contraction, and gathering up her shortie nightgown, grasp her tightening abdomen, and run round and round the birthing table yelling at the top of her lungs, "Lordy, Oh Lordy, Jesus help me!" There were many variations on this theme, including wordy and profane imprecations against the hapless male who had so unwisely impregnated her some nine months previously. This scenario would indeed make a memorable scene for some yet unmade movie.

I recall waiting the summons of the delivery room nurse many a night while this was taking place, usually asleep on the metal cart awaiting beside the door for the transport of a new labor patient, or indiscriminately to carry a body to the morgue if need be. All of this trained me to enjoy sleeping habits, which in later life allowed me to take a telephone call from labor and delivery room at home, dress, get in the car, and make it to the labor room in twelve minutes without ever waking.

During the whole year at the Grady, I never entered the operating room, nor did I ever attend a postmortem. It was a year devoted to "scut work" by cheap slave labor. I was glad to make my application to return to Vanderbilt for training in obstetrics and gynecology. I was accepted and did not read the fine print, or somebody slipped something over on us. Heretofore the second year intern had received room, board and ten dollars a month salary. The year I returned, Vanderbilt Hospital went on an economy drive, and cut out the $10. Now this doesn't seem much, but without this edge, one doesn't even have cash to buy toothpaste. When the orderly stepped on my glasses which had slipped under my bed, I didn't have enough to buy a new pair. Christmas that year included a pint of whiskey from the chief to each of the house staff. I was glad to sell mine to a fellow inmate for three dollars, which was all he had.

Whenever blood was needed on a patient, we rushed to cross-match our own blood, running the required laboratory work, including an "Eagle" test for syphilis. This meant sticking our own arm to draw blood. It's not as easy as it sounds. If we did not crossmatch, then we informed our nearest friend who had the requisite type. All this to obtain the $25 which was the going price of a pint of blood in 1940. This was our only source of money, for "moonlighting was not allowed," even if the time and energy were available.

We had an evening meal at midnight. It consisted of two slices of bread with whatever meat had been served at noon. No catsup, butter, or other spread on the bread. The milk was obtained from a five gallon can of milk set on the table, with resultant much spilling over the table and floor. The situation was so bad that the Public Health Department closed down the area. The house system of those years was based on a near slavery, with thirty-six hours on duty and twelve off.

I do feel that the practice of obstetrics was better then than now. We were much cleaner and observed aseptic technique in a rigid manner. This was before the advent of the antibiotics. The members of the house staff taking care of the infected cases on OB were isolated from "clean" obstetrics. We were not even allowed to eat with them, but were segregated in a different dining room and sleeping quarters. We were much cleaner in the days before penicillin.

There was not any of this nonsense about doing Cesarean sections frequently. All hospitals were inspected and if the C-section rate was over 4% the hospital lost it's accreditation. The University of Georgia in Augusta kept it's rate at one percent for years, and the mortality and morbidity rates were the equal of any in the world. The present 25% rate of most hospitals is directly due to the fear of getting sued. No physician has been sued for doing a Cesarean but many have been sued for failing to perform one.

One of our unenviable jobs was to insert a Miller Abbott tube in the post operative distended patients. This involved sticking a small (the size of the tube depended on whether one was the inserter or the insertee) tube through the nostril of the patient into the stomach. The double lumen tube had a rubber bag on the tip which was filled with a small quantity of liquid mercury. This added weight carried the tube down through the entire gastrointestinal track, and the suction applied to the other lumen rid the suffering patient of gas. (Occasionally the mercury laden tube came all the way through the GI track during the night. It was always a controversy as to which direction the tube should be most appropriately removed.)

On one such occasion, we had run completely out of the rubber bags needed to contain the mercury. I called the drug store and had them send over a half dozen condoms (for that was the most sturdy of the containers available). I told Doctor Henrietta Freund to go to the front door of the hospital, pick up the package, and hasten back for the lack of the rubber bag was delaying the insertion of the tube. Henrietta in her naivete, met the messenger and said, "Thank goodness you are here. We are in a terrible hurry for these." The good Pharmacist could hardly contain his curiosity, but hesitated to ask questions in such a delicate situation. Finally his curiosity got the better hand. He called to find out just what kind of emergency needed condoms so quickly. I got the impression that he was a trifle disappointed in the explanation.

On one of our infrequent trips to Percy Warner Park (still our favorite recreation) Gale and I noticed an airways beacon. In the old days the US Mail was flown by open cockpit airplanes and at night the course was plotted by lighted beacons. These were up about forty feet in the air, so we decided to climb one. We were on the small platform a few minutes and then decided to come down since it was beginning to get dark. The descending darkness triggered the lighting mechanism and the huge searchlight began to rotate, nearly sweeping us off the platform which had no railings. This was undoubtedly the most dangerous and foolhardy thing we did during our courting days.

Sometime during the spring of 1941, I acquired a "bug" and was hospitalized several days. I ran a very high fever every day, up to 106 degrees on one occasion. Although numerous test were done, the most astute diagnosticians of the medical school never made a diagnosis I was sent home to Spartanburg for a week of rest. During the hospitalization, Gale visited twice a day and her massaging of the severe back muscle spasms which accompanied the high fever was the highlight of the long day. It was during this trying time that she got the only low grade on a test that she ever experienced. Fortunately the calculus professor knew her well, and as he handed her the paper, he said quietly "Too much Doctor. Take it over."

The use of anaesthesia was included in our obstetrical training, and since nitrous oxide (also known as "laughing gas") was used (and still is) we were taught how to administer it. As a guinea pig, I volunteered, until I found out later that the fellow resident was proving to the class just how blue the patient could be taken and still come back, apparently unharmed. (The blue color was due to lack of oxygen.) Perhaps this episode could explain the forgetfulness of my older age.

Even though every effort is made to avoid mistakes, especially in the teaching institutes of medicine, mistakes occasionally happen. The human factor is always present. During my tenure on the wards, an episode occurred which depressed us all and made us keenly aware of the unexpectedness of fatal error.

One of the town physicians ordered a dose of strychnine to be given to an older patient. In view of the age of the patient and his lower weight, the Doctor wisely ordered a smaller dose ie. instead of the usual 1/100th of a grain to be given intramuscularly, he ordered 1/200th. The nurse, having only tablets of 1/100 grains, gave the patient two tablets of 1/100th grain (four times the ordered dose). The patient died as a result. There was no attempt at a cover up. The nurse was properly disciplined, the family was appeased by out of court settlement, and little publicity appeared. The entire hospital community was depressed however, and discipline was tightened.

The international situation in the late 1930's continued to decline and many of us who held ROTC commissions began to receive beckoning letters. On 1 May 1941, a directive arrived for active duty to begin 1 June 1941. and I was ordered to report to the Station Hospital, Fort Oglethorpe, Georgia for a physical exam. I managed to convince the powers that be at the time that several competent physicians were available in Nashville. Finally I was allowed to take the routine physical in the town where I was working. This rather stupid interchange of letters was a forerunner of many such orders in the future.

Consequently, on 1 June 1941 there appeared on the doorstep of the Station Hospital, Camp Nathan Bedford Forrest (named after the famed Confederate cavalry leader who gave the damn yankees so much trouble during the "late unpleasantness between the States") a really green First Lieutenant of the U.S. Army Medical Corps. At that, I was way ahead of the majority of the new army officers, for at least, I knew how to salute properly, to march in column, and perform an "about face" without falling all over my own feet. Perhaps that was why I was assigned at first to the Admission Office. That way I was the greeter of the even greener troops of the Chicago, Illinois, National Guard Unit ineptly designated as the 33rd Division. They were such misfits they were never, even after years of stateside training, allowed to leave the confines of the United States.

The town of Tullahoma was a forgotten little frontier village of about two thousand souls, counting the livestock within the boundaries of the town limits, half way between the town of Murfreesboro and Chattanooga, Tennessee. This innocent little village had dumped into it's unsuspecting vicinity some two divisions, the camp complement, and all the civilian workers concomitant with the training and welfare of some 62,000 individuals. Finding a place to eat, sleep, and sustain all of life's functions was a major concern.

Thus ended the frenetic pace of a Vanderbilt Hospital Resident as our "hero" enters the calm and soothing environment of the womb of the peace time United States Armed Forces.

54th Field Hospital - 79th Division
Major W. C. Herbert, Captain De Lucca and Captain Lynn Fredrickson

PART THREE

THE WAR YEARS 1941 - 1946

WAR IS HELL "I am tired and sick of war. It's glory is all moonshine. It is only those who have neither fired a shot nor heard the shrieks and groans of the wounded who cry aloud for blood, more vengeance, more desolation. War is hell."

– William Tecumseh Sherman

Caduceus

Staff of Hermes (Mercury)

The winged wand, Caduceus, is used as the symbol of the Medical Corps of the United States Armed Services. It was the wand of the Greek God Hermes (Mercury). Greek Myth tells the story that Hermes separated two fighting snakes with his winged wand and they entwined themselves around the staff. The Medical personnel during the Civil War frequently stopped the fighting in order to rescue the wounded from the battlefield, and therefore the symbol as used by the Medical service has significance. It had no previous medical meaning. The use of the Caduceus as a symbol of medicine is to be deplored.

Staff of Aesculapius

On the other hand, the wand with one snake entwined was the staff of Aesculapius, the Greek God of Medicine for centuries. All the early Greek hospitals were known as Aesculapiads. The American Medical Association has as it's logo the Staff with a single snake entwined and so do most knowledgeable medical groups.

In the early years of the Forties, only an idiot could not have foreseen an eventual confrontation with that embodiment of Evil, namely Hitler and his Nazi War Machine. Not being a complete idiot, (just a gullible one) and being promised that at the end of a year of training in the Army, I would be allowed to return to civilian life, I entered the armed services as a First Lieutenant. I had been a Second Lieutenant upon graduation from Wofford in 1935, and by virtue of a M.D. Degree, had been advanced to a First Lieutenant. With it's concomitant advance in salary, Gale and I felt wealthy, and agreed to get married soon. This after three years of courtship.

I've tried to recall just how much the First Lieutenant's salary was but cannot: somewhere in the range of $200 a month, I think. At any rate, we allotted $10 a week for groceries and life's amenities such as toothpaste, scrub brushes, underwear, etc. Sometimes there was enough left over for us to go to the Base moving picture show, which was exorbitant for it cost 25 cents each.

I get ahead of myself. I was just finishing up the year at Vanderbilt, (my second year out of Medical school), and was accepted to the Residency at Spartanburg General when the Army beckoned. Since I had been commissioned on graduating from Wofford, I really did not have much choice as to whether I would go into the Army. I could have deferred another year, I suppose.

The date of 31 May 1941 is designated as the time entering active duty, but I was lucky in that I was able to claim longevity pay dating back to the date of entering the Officers' Reserve Corps as of 6 September 1935, and separated from active service 20 January 1946. I can still recall the old tag number, O-335916. My base pay as a Major, including longevity was $287.58. I was ordered to report for duty on 1 June 1941, which I did.

A word of explanation is needed to those of you who might not be acquainted with the metropolis to which I was sent. The town of Tullahoma, Tennessee, consisted of a main street along the railroad from Nashville, Murphreesboro, to Chattanooga. The inhabitants numbered about 2500 persons. The Army in its omnipotence moved in the Headquarters and two Divisions. This totaled about 63,000 troops. In the three years we spent there we saw little of the camp, other than the Hospital. the Commissary, and the nearest Theater.

One of the Divisions sent to Camp Forrest, (named for the dreaded Confederate Cavalry leader Nathan Bedford Forrest), was the infamous 33rd Division out of Chicago, Illinois. This was a division composed of political bosses made officers, but by no means gentlemen, and the scum of the Chicago underworld. These men were sent to the wilds of Tennessee to be taught the fundamentals of soldiering. I'm sure attempts were made to instill some basics of discipline, but we in the base hospital never found any evidence of it. The Division was sent to various camps over the United States during the war and stayed in each location until the local authorities demanded its immediate removal, but never, never was it allowed to leave the confines of the United States. I have no doubt that Chicago wished that the men had never been allowed to return home.

To illustrate the lousy military attitude of the average soldier of the infamous 33rd Division, let me recall a chance encounter I had with one of the Chicagoans. I had charge of the Admitting Office when I first entered the service. This was a busy place and in contact with the variety of the camp stragglers. (Some seventy patients came through every day. During Maneuvers over two hundred was the daily quotient.) When I entered the office one morning, there was one of the Privates from the 33rd. He was lounging on the couch in my office and did not bother to even stand up. I yelled "Attention!" and he reluctantly stood, upon which I gave him a lecture on military courtesy, after I got him to attention. I kept him at attention the whole time. As a conclusion, I told him that if need be I would not hesitate to throw him in the Guardhouse for insubordination. By that time, the Admitting office staff was outside the door silently applauding for they had little use for the Damn yankees in camp. As the tough left, I heard a long drawn out "Cheeezzzz!"

On 13 August 1941, was issued the single most important order ever to effect my life. This order from the Headquarters Camp Forrest, read as follows, " 9. Under the provisions 605-116, leave of absence of ten (10) days, effective on or about August 18, 1941 is granted First Lieutenant William C. Herbert, O-335916, Medical Corps." And thus began the most exciting and continuing adventure of my entire life, that of marrying Kitty Gale Richards.

On the evening of the 17th, with one of my friends covering for me, my brother Elmore, and I set out, driving a Ford which I had recently purchased from my brother Thad. We only stopped briefly in Nashville to pick up Ethel Mary Gaskins, one of the bridesmaids to be. We drove night and day until we arrived in Oklahoma City. My memories of this city consist of "goose necked" oil wells pumping continuously night and day, surrounded by a purple haze over the sky and hot dust pervading all. It was so hot the populace never went to bed until after eleven and the Richards Family slept outside in the hedge enclosed garden at times. This was before air conditioning, as you have already guessed.

I recall a party that evening, but it's pretty hazy, since we had had no sleep. I do recall the "high noon" wedding. Everything was in white, including my white dress uniform. The newspaper photographer got a good picture of us as we left the church. Dad stood in for me as best man, and teased me at the altar, pretending to have lost the ring.

My only other vivid memory was that of trying to get the Gill brothers out of the car so that Gale and I could begin our honeymoon trip back to Camp Forrest, Tennessee. Since each of the Gill boys were six feet two, and weighted more than 200, it must have seemed funny to the relatives assembled for the reception, but I was not the least amused. At that time, I was six feet tall, weighed 145 pounds and had a 28 inch waist, no equal in any way to the burly ranchers and oil well owners of Gale's family. Incidentally most of them were graduates of Yale and Harvard in engineering.

We drove all that afternoon until we reached Fort Smith, Arkansas. I had reserved a room, and ordered flowers at the large hotel, but when we arrived a convention had taken all the rooms. Fortunately, as we were eating supper, someone told us of a new motel on the edge of town. Neither of us had had any experience with motels, but the Gods of Serendipity stepped in again. (This has happened so often during our lives: Something better will turn up. And so it did.) Our beautiful little cottage had never been occupied previously. We never missed the flowers.

The next day we drove on into a small town in Tennessee, and the heretofore faithful car quit. It seemed that the generator had expired, and a replacement had to be found. Again a small motel appeared for the night. The mechanic located and replaced the part. However this expense had not been anticipated. One of our wedding gifts had been ten silver dollars in a leather bag. This muchly appreciated gift was used to pay the bill.

The next night we arrived at Brevard Hotel, a large rambling old building, with antique heavy furniture. The bed was large, with embossed and engraved walnut carvings. It had originally had rope supports I'm sure. But we made no complaints, though as I recall it did make lots of noise. The following day we spent on an island above the falls outside the town of Brevard. It was subsequently a girls' Camp, and is now a housing development. It was quite deserted in those days. We made camp, blew up the air mattresses, ate, took a shower in the falls, and retired on an island in the middle of the stream with the sound of the falls to lull us to sleep. The following day we went on to Spartanburg, and then to Shelbyville.

Being in the peacetime army was a delightful way to spend a first year of married life. Originally Gale and I had an apartment in Shelbyville, Tennessee. This was a little town about fifteen miles from Tullahoma, and the war had made little impact on the sleepy village. The house was a large colonial home with the conventional four columns on the front porch, fourteen foot ceilings on the first floor, with an impressive staircase in the center hall.

The building had been divided into four apartments, two upstairs and two downstairs. We had the one on the right as one entered the front door. Our living room had been the parlor, our bed room the dining room. Our kitchen and bath were squeezed into what had been a butler's pantry. Across the hall was a couple from New Orleans, a Lieutenant Montagnet and his wife. I can't recall the two upstairs tenants. The four officers combined transportation, and it worked out rather well. Gale and I still had the little Ford which I had bought from my brother, Thad. The same one Elmore and I had driven to Oklahoma for the wedding.

Gale and I had never even seen the apartment when we drove in one night after the honeymoon trip from Oklahoma City. As we drove into town at sundown, the lights all came on and banners everywhere indicated a Horse Show was to be held that night. Gale remarked that as much as she enjoyed horse shows she would be too tired to go even if it was in her back yard (and that was exactly where it was: immediately behind the rather stately old Colonial house).

I do remember talking to the owner the next day, and telling him we only had the rent money for two weeks, (it was to be rented monthly), and the rest would have to await our next paycheck. I don't recall any hassle. The folks were very kind to us. Gale and I went back some fifty years later, and the house was still there, much nearer the race track, and much run down. (But so were we.)

It was in this apartment that I again came down with my mysterious illness, consisting of high fever, muscles aching. and on this occasion, delirium. Gale recounts that I repeatedly asked that my head (which I thought was in the upper right hand corner of the room) be returned to it's proper place. This must have been very frightening to a bride of less than a month. The illness must have been very short lived however, for I did not miss much work.

Another episode deserves mentioning. Gale took me to work at the Admitting Office of the hospital one morning. She braked at the Sentry Gate of the Camp. The Guard asked if she needed a Visitors Pass. "I don't think I need one, I'm just going up to the hospital." (It was clearly visible off to the right) "Well, if you're just going up there and let your father out you won't need one." This to newlyweds of a month or so. I was 27 at the time and Gale 23. This assumed paternal relationship continued and got worse as she got younger and my hair continued to get whiter.

The fall of 1941 was a delightful vacation. The hours were set, the weather was perfect, the living was grand. We often took picnics to the surrounding country. I recall many water falls in the area. We took along a .25 caliber pistol which I had gotten from Mother. I even had a medal for firing a .45 caliber during my camp days in Edgewood Arsenal, as well as an Expert Medal in Mortar Firing from the same source. (Chemical Warfare School-CWS.) Gale became a good shot with the little pistol.

We had gotten the pistol because of a scare. In our frugality we had saved enough to buy a War Bond, and on returning to the apartment we placed it on a table near an open window. In our trip to the bank to purchase the bond, some crook, probably from Chicago, had followed us, and learned where we lived. That night I was awakened by the sound of the man cutting the screen wire near my bed. I roused, (I have always been a very light sleeper) called out, turned on the light and the visitor fled. How he thought he could cash a War Bond I don't know. That was one of my many poor investments during my life. By the time the bonds matured, the money had deteriorated due to inflation, the governent demanded income tax so that we lost money on all our bonds. So what's new with the government?

The hours of the peace time army were really a treasure trove for newlyweds. Many days were spent all morning in the Operating Room. I counted one year that I personally did over three hundred hernias, which is more than many Doctors perform in a lifetime. After lunch, we frequently would match pennies to see who would be the backup for the OD (Officer of The Day). The rest of us then took off until the next morning, I tried golf one time, but being an ardent newlywed, the charms of the verdant greens and the rolling hills could not compare with ardent greetings of my new bride. Her hills and valleys and unexplored nooks left the Tennessee flora and fauna to be discovered much much later, and after fifty years they still come in a distant second.

That method of allocation of duties at the hospital left much work upon the young shoulders of the Officer of the Day. But before I describe the problems an OD faced, let me describe the way the Army builds an hospital in the "jungle" of the southern United States. First the engineers pick out a place as far away from civilization, as we know it, which is possible to find in the confines of the continental United States. It comes already well equipped with mosquitoes, flies, ticks, copperheads, poison ivy, chiggers, and in most swamps, full of crawly creatures that bite, sting, itch, and animals that "go 'bump' in the night." Things that the newly arrived unwilling visitors from above the Mason Dixon line had only read about in books, and which became terrifying monsters in the night when one was assigned to guard duty alone with an overactive imagination as the only companion.

Having found such a place, the engineers carefully add the all pervading dust of bulldozers gashing out roads, the knee-deep mud of poorly diverted streams, and a grid plan which resembles the best designed labyrinths of the European Gardens, guaranteed to lose the unwary within the first quarter mile. I am quite sure that if a conscientious effort were made to comb the area of the Former Camp Forrest, several soldiers now listed as MIAs of WWII would turn up in the hills of that God forsaken area. In my studies in the writing of the History of Medicine in Spartanburg county I found that the making of base hospitals was perfected during WWI. The plan was the same some forty years later.

The Hospital is always tucked away in a corner, entered through a hidden gate off a long country lane and guarded by a Prisoner of War compound. The guards of the compound are by necessity made up of rejects who have proven unfit for other army service, including KP and latrine orderly. The only other prerequisite was a love of firing the rifle or pistol on any and all occasions.

The placement of the new hospital is always the same. A rise is selected. The theodolite for surveying set up, and a quarter of a mile away is marked to indicate the end of the first bank of wards. This point is also on a rise. In between are numerous small hills and valleys. The instrument is then moved to the right 100 feet and the procedure duplicated. In the case of the Camp Forrest Hospital there were three such lines of ward buildings. These were then connected by a central corridor. Now instead of leveling the site with dozers, the foundations of the wards, consisting of brick pillars, were inserted by local laborers, who may or may not have encountered a plumb bob at some time in their lives and who never saw a level.

After the pillars had hardened the floors were laid, consisting of the newly cut lumber from the surrounding countryside. There was no time for drying of the wood. When the last hammer blow was struck, the carpenter knew the nail was set when the sound was a "splat" as the juicy sap flew everywhere. The wards were set above ground to provide a good home for the local fauna, such as snakes, opossum, rats, and a variety of vermin not even in the books of biology. This also allowed the winter's cold winds to circulate freely under the floorboards. This was to "cure" the wood over time.

Next the side walls were set up. Insulation? It hadn't been invented yet, and the new fresh lumber warped enough to encourage the least breeze to chase out the "bad" air and replace it with cold invigorating "good air." Perhaps this helped keep down the incidence of Tuberculosis since that was all we had to treat the "White Plague." Come to think of it, there were few cases of TB discovered.

The roof was next. Again no insulation, and the oppressive and all pervading summer sun did a superb job of heating the wards even though the windows could be raised with the aid of a jack hammer and one of the local King Kongs.

There was a front door to the ward, and a back door which was kept locked theoretically. This door was provided to facilitate the "borrowing" between wards. A note of explanation is needed. When the frequent inspectors arrived, Ward A1 was missing five sheets, ten knives, two forks, one bucket, and a plumbers plunger. These items were quickly borrowed from the Ward B1 immediately behind Ward A1, and when the inspector had been appeased, they were returned with any items needed to complete the inventory of Ward B1. One absolute necessity for each ward was an enlisted man who was trained and competent in all branches of larceny with an added degree in the art of "cover up." This was of much more importance to the running of a "good Hospital" than having competent Doctors or good Nurses. Excellence in thievery always rose to the top. The best of the practitioners of larceny became Master Sergeants. (A TV series similar to *M.A.S.H.* should be assembled lauding the accomplishments of the Master Sergeants during WWII. But more of them later in this epistle.)

On the left of the front door was the office, and across was the orderly's storage. The latrines were next. I recall coming into the ward one morning after Surgery, and being struck by the absolute silence. "Where is everybody?" I asked in a loud voice to myself. The Nurse appeared, entering with a loud "Shushhhh! Walter Pidgeon is on the Ward!" I demanded "Who in the hell is Walter Pidgeon?" At which time a tall distinguished figure appeared in the doorway. He and I introduced ourselves, at which time I remarked belatedly, "Oh yes, you and Greta Garbo were together in the movie *Miss Minever* recently." Maybe his ego needed deflating! Or perhaps I had been out of touch with the real world while in medical school.

The wards were pleasantly full most of the time, except during the time of maneuvers, at which time they were full to overflowing. These were always held in mid summer, and Tennessee summers are brutal. The insects and the heat made everybody miserable. One of the illnesses seldom encountered in civilian life was one called balanitis. As long as men are reasonably clean this does not usually appear, but when men are hot and unbathed for several days at a time, the uncircumcised male frequently gets infection under the foreskin. This causes the aforementioned disease. While not life threatening it is decidedly uncomfortable, and gives rise to imaginings involving impotence and permanent deformity, as well as being painful.

These fellows were usually confined to one ward and the maneuvers produced numbers of cases. Now, the circumcision of a newborn is very minor surgery, the pain lasting only 15 to 30 seconds. We used to give the newborn a cube of sugar wrapped in gauze and dipped in a good whiskey as anaesthetic. It worked. Obviously that was not sufficient for a husky young male. They had to be put to sleep, and either a circumcision performed, or if the infection under the foreskin was too severe only a dorsal slit was made to relieve the pressure.

Post operatively they did fine, however some malicious Chief Nurse always seemed to take great pleasure in assigning the most luscious of the bevy of nurses to this ward. When the lovely blonde bombshell appeared at seven a.m. the reaction was organic and automatic. As the nurse made her rounds between the two rows, the lovely cleavage caused arousal which resulted in two rows of small tents of sheeting. It was hot summer time. And when she bent over to adjust a bed or pick up a chart, the tight derriere caused sutures to part all down the line. It was finally agreed to assign the sexy cause of the hitherto unreported medical problem to another area of the hospital. The battleaxe who replaced her caused no arousal whatever.

As mentioned previously, the halls between the wards wandered up and down the whole length of the ward row. In keeping with the army policy of separating male and female as much as possible, the women's ward was at the extreme end of the corridor. However, the Delivery Rooms were in the center, a part of surgery which seemed to make sense. The medical services of the town of Tullahoma were overwhelmed, so all the wives and girlfriends came to the Camp Hospital. The girls labored in the Female End Ward, but as time for birth approached the patient was placed on a gurney and wheeled down the long hall to be delivered.

One night as delivery of one of the patients became increasingly close, the nurse and I put the patient on a gurney and with the help of two corpsmen we started down the long hall. Suddenly a hard contraction claimed the attention of the patient. She straightened out, I grabbed the head and shoulders, the nurse held the right hand and the two corpsmen held the feet. As was to be expected, the gurney obeying the Laws of Gravity sped merrily down the long sloping hall carrying both the top and the lower sheet, leaving the laboring patient clad only in the "shortie gown" which opened as usual in the back.

The errant gurney was eventually captured, the patient transported to the Delivery Room and everything was accomplished without further incident. On the way back, I mentioned to the corpsman that he would have something to write home about that night. He replied in an abashed tone, "Sir, I wouldn't dare. I signed up to go into the Air Corps."

The idyllic life in Shelbyville only lasted about three months, and the traveling daily to Camp Forrest along narrow Tennessee roads in the winter was a constant reminder that I needed to be closer to work. Therefore we began a search for lodgings in Tullahoma. It was not an easy task, for residents even rented out chicken houses. (Perhaps that is why Tennessee became the center of psittacosis in the US during WWII. Psittacosis is a disease of the lungs acquired by breathing the dust of droppings of parrots and fowls in general.)

One evening while still living in Shelbyville, Gale and I thought we should extend our hospitality to our hospital friends, so we invited Drs. Eddie Fox and Chrisman for dinner. Gale's Uncle Frank had invited us to go to a local hardware store to pick out something as a wedding gift. We took advantage of the offer, and selected a very nice waffle iron, which cooked two waffles at the same time. (The iron lasted until all our children were grown, so it was a good one.) Gale and I planned a menu which was easy to prepare, but we forgot one essential fact. Doctors are always late.

In accompaniment to the waffles and syrup, we chose little sausages. For some reason Gale also chose to make biscuits. Now we know only an accomplished cook can make Southern biscuits. None of this prepared dough in a frozen wrap was available in those days. Biscuits were made from scratch. Gale's first biscuits were about the size of a Confederate Minnie ball, 80 to 90 calibre, and so hard we all agreed that if there were to be an ammunition shortage, those biscuits could well qualify as an acceptable substitute. And by the time our late, late, late guests arrived the little sausages could also serve as lethal weapons. As I recall the syrup and waffles were a tremendous success since everybody was starving by that time. (The biscuits were not needed.)

Another of our acquaintances at the time was Doctor Baker Hubbard, a Vanderbilt Resident a little older than I. We were invited to play bridge one evening. Now I had been brought up on Wofford Campus and fully instilled in the evils of playing cards, but my bride insisted that I learn the fundamentals which included the names of the cards. Neither of us knew that Baker Hubbard was a card fanatic. At the gathering, the games began. At about eleven o'clock it dawned on me that the addicts planned to play most of the night, and eleven was past my bedtime. So when it came my time to bid, I bid six no trumps. I had nothing in my hand. I mean nothing. Gale was furious but did not show it. By superb playing and two or more finesses, she won the hand. To this day she has never forgiven me nor have we ever tried to play bridge again. Years later we have come to realize that Gale cheats. She has among her many other qualities, the somewhat scary one of Extra Sensory Perception, and somehow she and our son John have this in a marked degree. (This probably accounts for the fact that all her finesses worked.)

One summer Dad and Mother were with us at the beach. He was head of the Education Department at Wofford and had taught psychology for many years so he knew of the work of Doctor Ryan of Duke.

He placed John in one room and Gale in another, with John turning over the shuffled deck of cards. Gale, in the other room, caught the ESP signals from John and correctly called 24 cards straight without an error, as he turned them.

As a freshman at Wofford John won $60 at poker one night. When he told us, I convinced him that was, in essence, cheating. He reported later that he had tried to lose the money the following night but had won $35 more.

Moral: Don't play cards with Gale or John.

We moved to Tullahoma without difficulty since the apartments were "furnished," after a fashion. In fact, the only thing in the furniture line we had acquired was a marvelous record player that could play a whole stack of records one after another. On one occasion, the machine balked and I spent the Sunday afternoon taking the recalcitrant box apart. Gale was very supportive and said nothing, but one could tell that she was also dubious. Fortunately, all the parts went back together with none left over, and she has never again questioned my attempts to repair the various needs of the house and car. She's a goooood helpmate.

In our army the second Lieutenant was a "shavetail." Do any of you know why? The reason was economic. All officers have to buy their own uniforms, while the enlisted men have the uniforms furnished free (except they had to turn in an old one in order to get a new one, but that is easy for a scrounger). The second Lieutenant has a very limited income, so he goes to the Supply Sergeant, about whom a TV series similar to *M.A.S.H.* should have been produced. There he buys the needed clothes and his talented wife cuts off the rear of the shirt enough to make a pair of epaulets, sews them on the shoulders, and "voilá," the proud officer has his everyday work uniform. Gale modified most of my uniforms and many of my friends' clothes also. She has always made her own clothes through the years, and so beautifully that couturiers once followed her in Paris. The next year, Cape and Skirt came out copying her outfit.

But back to the duties of being an Officer of the Day. So far as I can recall these duties were allocated fairly. The only officer who did not pull OD was the Commanding Officer. Our first CO was a regular army medical officer who never interfered with any of the medical problems. Later Colonel Ferenbaugh was promoted to CO. Previously he had been a civilian doctor.

The OD's main duty was to make rounds after supper and take care of any medical emergencies that might arise before the morning shift arrived. Making rounds was no small task. Someone calculated the OD walked and rode seven miles during his tenure of duty. First duty was to visit each ward, tramping the corridors, and then to take a jeep ride around the perimeter. On one occasion I had finished rounds inside, and taking the jeep I went to the Admitting Office around eleven o'clock.

As I pulled up to the entrance, the sentry called "Halt." Since I was already halted, I said, "I am Halted." He said "Halt" again. "I am halted. Soldier, what are your orders?". (Which orders are "Advance and be recognized.") He answered, "Sir, I was told just to say "Halt" three times and then start shooting." We really had some weirdos in the medical rejects.

On only one occasion while I was OD was I truly apprehensive. The urologist and I had operated on a soldier, and the left kidney had to be removed. Ordinarily the renal artery can be exposed and ligated. This vessel is about the size of a lead pencil and can bleed excessively. Since this one was difficult, Major Daugherty applied a pedicle clamp to the artery and brought the instrument to the outside. At close of work, he went home as usual. In the middle of the night, the ward nurse called me saying the patient was bleeding too much. When I checked, she was right. Lucky for me there was still one ratchet remaining on the clamp, and when I tightened the clamp the bleeding stopped. I watched for a long, long time, and when no further bleeding occurred, I breathed a prayer of relief and went back to bed for the remainder of the night.

One of the times I recall feeling sorry for a lowly second Lieutenant was the time he was ordered to take a bunch of Medical Officers out to the field and teach them the basic drill. Most of the Doctors were between thirty and fifty years of age, had never before in their lives marched in formation, and cared nothing about doing so again. Looking back it does seem silly. Almost as silly as sending the whole batch of us to undergo the Infiltration and Gas Training Course. Can you imagine a batch of overage physicians crawling on their bellies with machine gun fire just over their heads, and explosions right and left of their flabby bodies. It happened. We trained vigorously with the 48th General at Memphis, taking 25 mile hikes and once we reached England, we never walked a step. We rode everywhere, even if it was on bicycle.

The Colonel wanted to have a big party for the Officers and Nurses of the hospital, and being in the middle of the Tennessee Square dance country he wanted a costume party, and a square dance. Gale and I decided to go as Daisy Mae and Lil Abner, two characters out of the funny paper of the time. They were two typical hillbilly types. When we arrived at the dance we found that aside from the famed Grand Ole Opera Star, Minnie Pearl and her partner, we were the only ones in the hospital personnel who knew how to square dance. The four of us spent the evening teaching the others the intricacies of hillbilly dancing. I recall that Gale was a lovely Daisy Mae.

We stayed in Shelbyville some three months, after which time an apartment opened up in Tullahoma, probably from some medical connection, for our landlady was the wife of the local Doctor. The old Victorian type house had simply been converted into two apartments by closing the double glass doors between the two sides, and putting curtains on the other side from us. A bathroom was added onto the porch by putting up studding and adding sheetrock on both sides. No evidence of insulation was ever found and the winter wind easily made the bathroom a place not to linger in the cold months.

Company entered our apartment through the front door leading into the living room. This room was seldom used except for visitors. To the left near the window was a large lounge chair which looked rather inviting. Whenever company came, it was my job to stand in front of this chair as master of the house. The aforementioned inviting chair was covered with a slipcover, obscuring a large hole. Only by sitting on the front edge could one avoid falling through the large round hole in the bottom.

The next room was the bedroom, which space was mainly occupied by a large Warm Morning heater. This, to the uninitiated, was a large steel cylinder, the size of a fifty five gallon drum, which could be filled with coal, and would burn slowly for a long time. I recall our going on leave for a week, and upon returning found the fire still glowing.

Occupying most of the rest of the room, was a large iron bed. One night Gale was fixing supper, and I was reading abed. When she called to me come to eat, I was mesmerized by some book and did not answer. Whereupon she came in, jumped on the end of the bed to get my attention, and the bed promptly collapsed. The giggles and shouts of laughter (I almost wrote Gales of laughter) penetrated the glass double doors. The girls who occupied the other apartment had placed a large sofa against the double doors, and one of them was having a date when the commotion occurred. The boy, intrigued by the laughter, said "I've just gotta look." To which his girlfriend replied "You can't. He's a Major and you are only a Sergeant."

The remainder of the furniture was what we called "Orange Crate Modern." The oranges were shipped in sturdy wooden boxes at that time. We covered these boxes with wallpaper or creton, and stacked one on another. These were quite acceptable substitutes for bookcases, end tables, and bedside tables. We rented furnished apartments, but believe you me, the furnishings were Spartan, including the oil-cloth covered kitchen table.

On one occasion as Gale's birthday approached, I found an old baby carriage in the garage. I asked our landlady if I could have it. I had expected her to give it to me, but no such luck. She charged me five dollars. I also cautioned her not to tell Gale about the transaction. With diligent cleaning, some paint, and a nice smooth plyboard top, it made a quite acceptable tea cart. Our landlady, on coming to tea one afternoon, confessed to Gale that she had wondered for a long while why she was not to tell that I had bought a baby carriage.

Beside us was one of the large Victorian houses of the town. One of the extremely wealthy Rockefellows had rented it. He always changed to civilian clothes when he came home from camp work. But what really set Gale's teeth on edge, was the frequency with which the chauffeur or the maid would pluck the flowers on our side of the fence, carefully tended by us, to put on the Master's dining table for the evening meal.

One of the delights of our days was to visit the Army Commissary. (Perhaps this is why I like to shop at Sam's so much.) Most of our food came from this source, for the local groceries were swamped. On one occasion Gale bought a nice plump chicken to have for dinner. When she tried to prepare it for cooking, she found it completely frozen. After struggling for some time, she found to her horror, that though the feathers had been removed, the 'innards' had not been cleaned out. One of the girls next door was of inestimable value to the new bride and they taught her how to "gutt" a bird. The girls were secretaries in the various offices throughout the camp. I think four of them occupied the apartment.

During my last year of Medical school, I had borrowed $300 from Uncle Roy. This was repaid when we began to save a bit. The check was sent to him sometime the last part of our first year of married life. More than thirty years later, I got a letter from the Treasury Department saying that War Bonds for the amount had matured and no longer drew interest. I had been made the designated beneficiary provided I could prove I was Dad's son. Since Dad was dead by that time I had to send his death certificate. Dear Uncle Roy had sent the money to him and he had placed it in the safest place he could find which was War Bonds.

7 December 1941 arrived. The statement has been made that certain times etch themselves on one's remembrance. This was certainly true of the "Day which shall live in Infamy." I was washing the car outside the apartment when Gale came running out to announce the news which had been on the radio: The Japanese had attacked Pearl Harbor! We knew then that there was no returning to civilian life until the evil which surrounded the whole world had been crushed. When I went in to the Hospital later that Sunday morning, the sneak attack was the sole topic of all conversation.

Soon after the declaration of war against Germany and Japan, (I can't even recall if Italy was even given the dignity of a formal declaration, can you?) the Vanderbilt unit was called into service and sent to Camp Forrest for training. About the time that I was made Captain. (1 January 1942).

One of the perks of the Medical Officer was to take patients to the various hospitals close to their homes. These were choice since we were paid per diem, meals and mileage. Except when the mileage was over railroads controlled by government grants (which were mostly the long trips to the west over land grant rights of way). My first trip was an easy one to Lawson General in Atlanta. These were never sightseeing tours for we were required to return to base upon completion of the task, but they did bring in more money.

We were required to attend numerous schools which were of value I'm sure. One of these was The School of Military Law which resulted in our being placed on military courts. Many of these trials were of individuals who were unsuitable for military service, (or considered themselves unsuitable). The great majority of the trials were on the basis of urinary incontinence. This assumes importance when the participants are housed in barracks with thirty other men and half are in the upper bunk. The ones in the lower bunks tend to be vociferous in their complaints. Standard military cures included putting two of the offenders in the same bunk, alternating nights as occupants of the upper bunk. If this did not succeed, the unhappy lad was put outside to sleep, preferably in cold weather, with only a sheet as company. If all failed he was brought as candidate for CDD (Certificate of Disability Discharge), in those days it amounted to almost a DD (Dishonorable Discharge).

In one court, consisting of three Medical Officers, I was the President. The lad before us for questioning was a lanky uneducated Tennessee hillbilly. Question: "It says here in the resumé, that you wet the bed every night. Is that true?" Answer: "Yes, sah! Every night." Q. "I see here that you are married. What does your wife say about your wetting the bed?" A. "Lawsy, Major suh, she wets the bed too."

After a brief consultation, the three members agreed that a man in a foxhole did not need the added complication of a mate bothered with enuresis. A member later queried "Wonder what that Tennessee bed of corn shucks smelled like following a few nights?"

On 30 September 1942 came the order creating me a major. Hot Digity Dog! That is the best rank in the armed service, for its high enough to be included in the field grade officers and too low to have any real responsibility. Along with this advance in grade came some suspicious duties. I was made Chief of Operating Room (OR) and Anaesthesia, and shortly after that Executive Officer of the Hospital, 2 November 1942, replacing Major Sol Goodman. These duties had a suspicious feel to them. And not too many months later these feelings were confirmed. I was offered the command of a field hospital being formed. This carried the rank of Lieutenant Colonel with a sure increase to Colonel eventually. Though highly flattered, I requested not to be assigned to this duty. One, I'm not sure I could be a good CO at my age 28, and the Vanderbilt Unit had recently been assigned to duty at the Tullahoma Hospital for training. These were all my contemporaries, and I knew them as top notch surgeons. Oddly enough they did not seem to resent the fact that I outranked many of them.

At one time Doctor Tommy Frist appeared for a physical to get into the army. Now Doctor Frist had been our medical chief at the Tennessee State Penitentiary two years before. I knew that he had had two spontaneous pneumothorax episodes previously so that under the regulations I had to deny his physical. Here am I refusing the chief. Fortunately he was able to get a waiver from higher up. He is now president of the largest hospital chain in the United States.

The year the Vanderbilt Unit spent with us was of inestimable value to me. I learned a great deal of surgical technique and diagnostic acumen, but most of all I acquired self confidence. I came to realize that I was a better surgeon than most. This was to come in very handy in Europe. And there serendipity played a part. Had I taken the CO job of a field hospital I would have done little surgery, for his duties are mainly administrative. And indeed I got to spend another year at home doing an immense amount of good surgery under good teachers. After the war I saw Colonel Ed Rank, who took the field hospital job and he had not been very happy in his assignment.

I was urged to take the Surgical Boards to become a Diplomate being told that "Everybody passes. All you need is $25 and a pint of whiskey." Which was true. But the Vanderbilt faculty had rather looked down on these certifiers, and very few of us applied. I still have the letter of indorsement in the 201 file. I know now that it was a mistake not to have taken them.

I recall one case in which the Colonel of the VU unit came to my rescue. A very nice lady was to be operated the next morning for a mass in the right groin. I had thought it was an inguinal hernia, but Colonel insisted that I look at her right foot. There was a black spot not more than the size of a pinhead. No other findings were present, but I was persuaded to send her to Vanderbilt without much difficulty. She indeed had a malignant melanoma which had metastasized and she was dead in a short while.

On 20 May 1942 I was appointed Assistant Chief of Profes-sional Services for the Station Hospital.

In case you are misled by my writing that it was all fun and games, I will list the number of operations done during the following months, February 1943 - 10 major procedures; March - 22; April - 21;

May - 14; June - 18; July - 31; August - 40; September - 45; October - 38; November - 10. I recall one count revealed I had done over 300 hernias in one year. This is more than the average surgeon performs in a lifetime. All this time I was perfecting the meticulous technique passed on by "Doctor Barney Brook's" residents. Doctor "Barney Brooks" was the Chief of Surgery at Vanderbilt, and aside from being an ogre who scared the daylights out of all the students, was the most marvelous teacher I ever encountered.

One of those perks which I mentioned before, involved taking a psychiatric patient to his home town of Lyons, N.J. The chap was medium size and quite amenable most of the time, but known to be a homicidal maniac. We assigned him the upper bunk on the train. I took the lower, one ward attendant slept on the other bunk, and one stayed awake during his sentry shift. Needless to say, there was no sleeping on this sentry duty, nor do I recall sleeping much. Everything went well until we had to change trains in Baltimore. When I went to the ticket office to confirm our tickets, the patient got away. Aided by the MPs, the local cops and everybody else who was informed of the Diagnosis of Homicidal Maniac, the poor fellow was soon recaptured. I'll have to remember that episode if ever I need assistance again. Never has such cooperation been commanded in such a short time. The rest of the trip was uneventful.

To add to the exciting adventures of the year of 1942, Gale got pregnant, with help. She had a pregnancy remarkably free of any difficulty, but I must admit that I had morning nausea on frequent occasions during the early pregnancy. This I feel was carrying the French idea of "couvade" entirely too far. This ancient custom, still carried on, particularly in Normandy, has the happy father go to bed after the birth of the new baby. The new mother gets up and prepares a feast, invites all the neighbors in to bring gifts to the father, while she slaves away in the kitchen. While I endorse the idea in principle, I have the feeling it will never catch on under the present NOW mystique regarding feminine demands. It did not engender too much enthusiasm from Gale when I mentioned it in the unenlightened years of World War II. If I were to make it a platform plank for the present day politician, it would ensure his overwhelming defeat.

Sometime during mid pregnancy, I was issued a gas mask. When I got home I put it on and crept silently to the kitchen, crawling on hands and knees. At the appropriate time I made a growling noise, and very properly scared her. Any postnatal defects in the progeny were always blamed on me, because of this prenatal influence.

Gale went into labor on the 16th of December. Now wouldn't you know that that was just the time that Captain Eddie Fox, her obstetrician had chosen for his Christmas holidays. As we sat there in the Women's Ward on a bleak 16th December day, I kept reassuring her that everything was progressing normally. I was reading some book, as usual, only this time out loud to distract the laboring patient from the pain of her early labor contractions. This is called "Vocal Anaesthesia," and this is the first time I recall using it. Late in the afternoon, Eddie stopped by before he went to supper.

He examined the patient, jumped up, and yelled at me. "Good Lord! You're sitting here and she's about to have the baby in bed." With much rushing we made it to the delivery suite. While they were having the excitement of the arrival of Bill III, I was busy delivering another in an adjacent room. A to and fro conversation kept me up to date with the progress in the adjacent room and I was informed of my son's arrival at the same time the patient I was attending delivered. I cannot really say whether she had a boy, a girl, quintuplets, or a two headed gorilla!

Doctor Fox got to go on his Christmas leave without further incident. He had an idea that he could determine the sex of the unborn babe by choosing the dominant parent. He refused to guess as to Gale and myself because he claimed two Chiefs had gotten married. He was a very astute observer to pick it up even in the early days of our twenties.

Sometime during the year we had the good fortune to learn about an apartment in the home of Mr. and Mrs. Addison Mitchell on the main street. For nearly two years they were the locum parentis for the two and later three of us. We were so impressed that we named the last child John Mitchell.

The room across from us was rented by an FBI agent. He was so proud of his beautiful new car. Being an agent entitled him to many things which the rest of us could not obtain, not the least of which was a set of new tires. In order to protect his love he had put every anti-theft device on his convertible that had been invented. He also parked the car underneath his open window. (Remember no air conditioning, all windows were left open during the summer. They were screened of course.) Great was his consternation one morning to find his car propped up, with all the wheels gone, and all his anti-theft devices still in effect. We were properly sympathetic, but also a bit amused. Gale had said she felt so safe having an FBI agent next door.

Bill III fitted in the household beautifully. The Mitchells loved to baby-sit him the few times we went to the picture show, or to the commissary. And without a doubt he was the smartest baby the world ever saw. At the age of six months he was doing trigonometry by unfolding the neat triangles into which his mother had folded his diapers, and he enjoyed the stories of Greek Mythology with which I read him to sleep.

And speaking of diapers, the ritual was to rinse the offending garment in the commode, after which it was placed in a bucket of chlorine water, and the following day it was washed in the bath tub and hung out to dry. Great was the rejoicing, when one of the divisions had to move out and I was able to obtain one of their washing machines. It was a round one with agitator and a ringer on the side. You have seen many of them sitting proudly on the front porch of the Tennessee Hillbilly shack. Well, ours was on the back porch. and drained into the vegetable garden. Ours worked well after I rebuilt the gears, and replaced the dryer cylinders. These, Gale said, had been used to run the boots through the wringer.

But truthfully, we slept on a high four poster bed, and frequently left the baby on the bed. Before he was six weeks old he had been taught to grasp the bedcovers and slide down the side to the floor feet first. The first time Gale came in and the baby was missing, it was a shock and relief to find him happily playing under the bed. We are both sure that this lesson contributed markedly to his discovery in later life that jumping out of airplanes could be deemed fun. (He went on to become Executive Officer of the world renowned Golden Knights, an Army group addicted to jumping out of perfectly good airplanes.)

Also at less than six weeks I hung him on the clothes line in order to test the prehensile strength of a newborn. He passed with flying colors.

But the idyllic country life was soon to come to an abrupt end, for the war was progressing and 1944 was fast approaching. On 8th

November, I was issued orders to report to the CO of the 48th General Hospital, being formed at Kennedy General, Memphis, Tenn. Thanks to friends at the Camp Forrest Hospital Gale was able to get an apartment on the ring road near the hospital for a month.

However I was able to leave the base very little, for our CO was determined to make a functioning General Hospital in a very brief time, and unlike most hospital units, ours was gathered from all over the South. Most General Hospitals were products of Medical School units.

We drilled and attended classes daily. Frequent marches out of town were the rule. I recall one twenty-five mile march (at least it was told us it was twenty-five miles, and by the end we all believed it). We hiked so often that we were beginning to call our unit the Jackson's Foot Calvary after Stonewall Jackson's famous Army of West Virginia. On one overnight bivouac the temperature fell precipitously. The next morning the respiratory water froze on the inside of the pup tents, and the canvas refused to fold up. Brother it's cold when a pup tent stands up by itself after the tent poles are removed.

Parting was not "such sweet sorrow." The our last night together is best expressed by Tchaikowsky's "Nur wie die sehnsuch kent."

At last the day of departure arrived. Our orders were secret and none were issued to us as individuals. We left Memphis so as to arrive at Camp Joyce Kilmer (which has no trees within it's entire circumference. Joyce Kilmer wrote the well known poem, *Trees*). Where we were placed in the camp I have no idea. The following day we were bussed across the river to the docks in New York, and ensconced in the magnificent liner, Isle de France. It was estimated that some 14,000 troops were aboard. On Christmas Eve, the liner proudly backed into the Hudson River on its maiden run after refitting as a troop ship. Immediately, the ship caught fire because some idiot had stacked the wool blankets between the smoke stacks. We were evacuated immediately and returned to Camp Joyce Kilmer. In the meantime the fire boats continued to pour water into the ship. On and on water was poured in until finally the unhappy liner turned over in the slip in the New York Harbour. And there it stayed until cut up for steel scrap.

Can you imagine the consternation of the young Lieutenant at Camp Joyce Kilmer on being informed that 14,000 guests were being sent back to your keeping, and you were to find food, lodging, sheets, and pillows for them immediately? That was the case, for all the other officers were at a big party in New York, it being Christmas Eve during war time. I spent that Christmas in a ditch of red clay listening to Christmas carols which were incessantly broadcast over the camp loudspeakers. Sometime during the wee small hours the harassed young officer found room for us. The meal was K rations. Seldom have I felt any more sorry for myself than I did that Christmas.

The Queen Elizabeth I arrived shortly after the above episode and the 48th General was squeezed in among the other 16,000 troops. I recall that there were eight officers in the forward stateroom on the port side high above the main deck. It was really a magnificent cabin except for the fact it had been designed for two occupants. As soon as the occupants had been installed one of us, a Doctor S— announced that he was seasick. I rebuked him with "You can't be seasick! This is the biggest ship in the whole world, and need I remind you that we are still tied to the dock in New York harbor." "I don't care" was his reply. "I'm so sick I can feel the boat rock whenever someone steps on or off the ship!" And so it was, for he retired to his bunk and did not move until he was taken off on a litter in Scotland. All during the five days of the crossing we fed him intravenously, and he looked awful that last day. My last remembrance of him was seeing a small speck being lowered over the side in a litter. We had landed in Grennock, Scotland on the Firth of Clyde, I believe. It was a typical grey overcast day, but even today I still can see in my mind's eye, from our position on the dockside, the tiny litter swinging out from the huge ocean liner. The crane was one of those gigantic machines with which tanks and large cargos are picked up effortlessly and lowered to the dock. For a few moments, our friend made a pitiful spot against the leaden sky before he was carried off to some hospital. He evidently went into a replacement depot for we did not ever hear from him again. How did he ever get home? I know it was not by boat.

Doctor Fredrickson recalls watching the scene with a British inspector at the Docks. One of the Loading Nets broke while over the cement landing zone. The contents were spilled everywhere, but dozens of workers came from everywhere to collect the debris. The inspector inferred that every tenth net spilled the contents upon the dock for the benefit of the union dock workers: this, during the desperate straits of England's survival.

The North Atlantic is no place to be in the winter even in the most luxurious and largest liner in the world. All those of us who were not seasick were thankful for the forty foot waves we encountered. We had been told that the U boats had been given special instructions to sink the Queen Elizabeth, and that a curve of Underwater craft was spread across our path out from Greenland. Each officer had to spend one watch out on the deck with particular emphasis on periscopes. Each of us stood watch fully clothed, and with steel helmet buckled on because of the wind. It has been said that the life expectancy of a man in those cold northern waters was five minutes, but we felt that the wearing of the helmet was to assure us of a quick death by drowning since it would pull anyone down immediately. We all thanked Gale each time the watch changed, for she had knitted a Balaclava helmet liner of wool for me. and the next watch man appropriated it. This helmet was close knit, warm, almost waterproof, and only exposed the eyes.

The only other watch we pulled was an eight hour shift with the enlisted men. The huge holds of the liner had been built up with bunks of steel so that twelve men could lie one above the other, and another rank beside it with just enough room for one man to pass. Opening the door was a real blast of stale, pungent air. Human beings packed together can conjure memories of the so called Black Hole of Calcutta! Our men had had a bath just before boarding, but it truly made no difference after a few hours in that hell hole. The only time they left the quarters was to have a meal twice a day. The kitchens worked 24 hours a day to feed the horde. I can't remember the meals so they must not have been too bad.

We were all on deck only one time for boat drill. It was a thrill, even in the cold wet windy day, to see that many men together so quiet and orderly. I imagine the ship authorities found it a good time to open all the doors and air out the joint.

We were efficiently hustled into a train at dockside, and off we went to Southern England. All through the night we rode until we reached London the next morning. There the cars were switched to another line and we were deposited without fanfare in Petworth, Southern England, a little village dominated by a huge house and grounds. Petworth House was and is one of those Treasure Houses of

Britain.

For our "comfort" we were installed in several English half-round Quonset huts. These were half sections of pipe, measuring fifteen by thirty feet, made of metal with wooden floor, minimal windows, and heated by a small (and I mean small) upright stove at one end. The fuel was a cheap grade of coal which smoked more than heated. On one occasion, some of our boys found a dead tree in the woods which had fallen down. They cut it up and had a warm fire for a little while, that is until the British warden found out about it. It seems all fallen wood belonged to some forest association formed several hundred years ago and was sacrosanct. I think the American taxpayer paid for it.

The only heated area in a British town was the picture show, The entertainment was often provided by the courting English couples who met in the warm theatre. The movie was incidental. Occasionally we would have the transportation to town. One night as we left the show and piled into the eight by ten truck to go back to the camp, a typical English fog rolled in. Though the driver had taken the road numerous times it was necessary to put a man on each fender to call out where the truck was in relationship to the middle line and the outer edge. Even with this precaution the driver missed the gate into the grounds and had to back up. Nobody can describe adequately how thick such a "pea souper" can be and how disoriented one can become in it.

The people were very kind to us, and we tried to reciprocate. The Lady of the Petworth House invited us on a tour of the mansion. Most of us were not sophisticated enough to appreciate what we were shown. I, for one, did not know how to appreciate the magnificent works of the master carver, Grinling Gibbons. The carvings of the staircase, ceilings, and bookcases seemed impossible to have been carved out of wood.

We had been in Petworth only a few weeks, when we were moved to Swindon, the armpit of England. I suppose we were moved just to give us an exposure to the various kinds of life in the country. As I recall, it snowed most of the time. As we left the train, we were taken by Bobbies two by two, marched up the street to private dwellings. The Bobby knocked on the door, spoke to the lady of the house, and left us standing there. She obviously did not know we were arriving that day. This was our first experience with "billeting." We were led upstairs to a very cold bedroom. The furnishings were nice, as was true throughout the house. In the seven room house, there was only one fire and that was in the living room. In our two weeks stay, we became friends with the family and shared our rations of cigarettes, (these became the currency of negotiations in time) and the children, of which there were two, were won over by candy bars. I don't think they had had candy before in their lives. However, with only one small fire for the entire house, we did not think that we should intrude on the family life too much.

We did not ask too many questions about what the man of the house did to make a living. These were war times and secrets were not shared readily. On one delightful occasion, the two of us officers were bundled into his small Austin and driven down to a nearby town where he had a business meeting. I now realize he had to be an influential citizen to have a nice house, a car and gasoline to run it. He took us to a pub, the first time for us, and after a plowman's lunch of bread, cheese, onion and "arf and arf" (half beer and half lemonade) we admired the furnishings of the pub while our host attended to his business. Across the serving board, in the parlor of the pub was a beautiful carving of the Battle of Agincourt. It was black with age but had been lovingly preserved. I would like to see this magnificent art work again.

It was at Swindon that I found out that chilblains was not just something out of Dicken's writings. Prolonged exposure to wet cold will interfere with the circulation of the feet, leaving the arteries in a spasm. The feet remain purple with red patches, and the victim has a chronic ache in the afflicted feet. I am told that the natives take branches of the gorse bushes and beat the skin of the feet and legs to restore the circulation. We did not do much better in our treatment. In order to avoid the disease, we spent most of our time in the mess tent or in the British theatre. This is how I remember the dirty, coal ridden, railroad center of Swindon, and neither the "Billetees nor the Billetors" cared much for the encounter.

It was here also that we experienced the first casualty of our outfit, other than the seasick Doctor. As mentioned, we were billeted in the homes. Baths had been few and far between, so when our Doctor Drennan entered a house with a nice bath he decided to indulge in a hot soak. Now over the tub in the English house even to this day, one can find a large round can which has a coil of copper tubing in its core. By lighting the gas jet, a thin stream of hot water will fall into the tub, and eventually one can achieve a warm bath. Unfortunately Doctor Drennan forgot to read the instructions on the side of the "geezer." These included the admonition to open the window when the gadget was in use. Perhaps the fact that it was snowing heavily outside had something to do with the window being left closed. At any rate when his fellow billeter, Captain Pat Ciaglia, returned Doctor Drennan had expired from carbon monoxide poisoning. This rude awakening to the realities of war shook all of us.

Our next journey was not too far from Swindon, to a farm area between Winchester and not far from Stockbridge, between Upper Wallop and leaving Middle Wallop and Nether Wallop below us toward the coast. (Ahhh. I knew you would know exactly where to find us.) Don't forget that all the signposts and directions had been pulled down, in anticipation of a German invasion, and left that way because of the American arrival. As they said many times here in England, "You cahn't miss it!" but don't count on it.

Winchester was for many hundreds of years the capital city of Britain. Therefore the real history of the Island began here and I for one enjoyed it thoroughly. I bought a bicycle for four pounds, (an extravagant twenty dollars) second hand, of course. Never has a purchase been put to such good use. As I have said before we never walked anywhere after we landed. We bummed a ride or rode the bike or took a train (these were always crowded), and this city, so crammed with the past, was only ten miles from the hospital. I gave frequent talks to the members of our hospital and arranged some of the trips. In fact I was called "Old Ruins" by some. Our first trip was to a castle ruin which had been the home of General Oglethorpe, founder of the colony of Georgia. Our next was to Arundel Castle, home of the Dukes of Norfolk.

The hospital itself was laid out much as the ones in the US, except for the fact that we had to go outdoors to get from one unit to the next. The English didn't seem to mind the rain as much as we did. The mess hall was near the entrance. The up hill and down hill grade was more than at Tullahoma. The prevailing wind was down from the mess hall and as we went to eat, the menu could be discussed after smelling the cooking odors. "Goat again!" I'm sure it wasn't goat really but it surely was a very old mutton. And the smell of "little cabbages" (Brussel Sprouts) was a sure fire way of killing any hearty appetite. These smell just like the cooking collards smell in the South: to be explicit, similar to the aroma of an outhouse on a summer day. These delicacies were the English return for our Lend Lease of canned ham and fruit cocktail. When I returned home I asked Gale never to have lamb, Brussels Sprouts, or oatmeal in our house ever!

We set up house in the wards which were similar to the wooden ones back home and awaited customers. The only patients we had resulted from the way the "Crazy Yanks" insisting on driving on the wrong side of the road.

For entertainment, we got on our bicycles whenever we heard that an American hospital ship was at Southampton. It was a mere twelve miles bicycling down to eat a dish of ice cream and a twelve mile cycle back, uphill this time, but it was worth it. We did miss ice cream. We had been informed that the English did not sterilize the milk and that the herds had bovine tuberculosis. Not wishing to be candidates for the "King's Evil," scrofula as bovine TB was called, we were careful not to partake of milk products. We did try to reconstitute the vile stuff sent over as "dried milk." We even added bacon drippings to replace the fat but nothing approached the taste of the milk we had remembered.

Before the influx of patients we were required to take short hikes as a unit throughout the countryside. There was little grousing, for the weather had become delightful. Though the total strength of the hospital unit was around three hundred, the marching members made up two companies. The in cadence singing was quite lovely. In a vain effort to inject some elements of "couth" I did try to insert some of the classic songs into the repertoire, such as the World War One favorite, *Mademoiselle from Armentieres, Parley Vous? Mademoiselle from Armentieres, Parlez Vous? Hadn't been kissed in fifty years! Hinky Dinky, Parlez Vous!*

This has numerous raunchy verses as well as being a musical treat for marching, but it did not replace the favorite, nor did my choice, since we were going into France, which was "Au pres de ma blonde, il fait bon dormir." We always sang loudly and clearly, "Roll me over in the clover, Yankee soldier, Roll me over, lay me down, and do it again!!" While this made for excellent marching music, it did not make for reassurance to the English mothers and fathers through whose village we happened to be tramping. In fact we were forbidden to stop as a group, and were not allowed to enter the pubs as a group, for we tended to drink up the entire monthly beer supply of the area in one sitting.

The evening hours were often quite dull in the camp. Too much sitting and watching the lovely evening twilight, or watching the Royal Air Force knock down the V I bombs (June through August 1943), or waiting for the routine call of the cookoo, tend to become monotonous after awhile. Of course we could always call on Major Daugherty, the doctor in charge of the medical wards, to entertain us with an "inspiring" (meaning consumption of one or two drams of scotch) rendition of the long Ballad about the war between Henry the Second of England and Charles of France. This was the poem that cost Rudyard Kipling the chance to become the Poet Laureate of England, a position he very much wished to attain. The line which muchly offended the Queen ended each verse with, "He tied a thong to the Royal Dong, and led him around the Camp." I have tried to find a copy of this epic poem, but thus far have not succeeded.

A short bicycle ride of five miles brought us to Nether Wallop, an unchanged English village, with Tudor houses, and a Norman Church. It was so small I don't even recall seeing a Pub. I went by myself to the Village one day, and having parked my bicycle inside the litch gate (that is the gate where they put the coffin to determine whether the candidate has led an exemplary life to entitle him or her as the case may be to enter the consecrated grounds to await the Resurrection). In those days, the possessions left were still there when one returned. In the church, I met a delightful older gentleman who told me the history of the church and village.

I introduced myself, and he was Sir Guy Campbell, brother of the famed racing driver, Sir Malcomb Campbell, at that time holder of the world's speed record with the Bluebird Auto. After a few minutes of conversation, he invited me to tea. We entered a typical thatched roof cottage, and in the overstuffed sitting room was a lovely painting over the mantle, obviously a Gainsborough. I admired the painting of the little girl. It was revealed that it was indeed a Gainsborough, and the subject was an ancestress of the owner. As the tea and conversation progressed, it turned out that my host was General Sir Guy Campbell of the British War Office.

He in his turn found out that I was of some distant kin to the present "Reggie" Herbert, present Duke of Pembroke, and that I held little hope of seeing the Herbert Home of Wilton House since the House had been preempted by the British Army for the planning of the invasion of Europe. However in a few days the postal clerk at the hospital looked me up with some awe for he had a letter to me from the Headquarters of the British War Office authorizing me to enter Wilton House and addressed to Sir Reginald Herbert, Earl of Pembroke.

At the first opportunity, I got on my bicycle and took off for Wilton House. I had passed near the place on a trip to Salisbury and on to Stonehenge, which trip ended up totalling sixty two miles, beginning after lunch and ending back at camp by supper time. This was in the times BG, before graffiti, and one could wander all through the ancient stones, coming up from the Heel Stone to the Altar stones, all alone, with no guide, and especially with no barbed wire to keep out the public.

I made it to Wilton in record time and approached the sentry at the gate which supports the metal statue of the mounted Marcus Aurelius, which had been copied from the original one on Michelangelo's Square on top of the Capitoline in Rome. The sentry was completely at ease, and was the only military presence I encountered. Remember, this was the Headquarters of the British Fifth Army, and where the Invasion was being planned. The sentry took my note from the General, called someone to take it to the proper authority, and in a short while an older gentleman, clad in somewhat seedy knickers, came down to the gate.

He introduced himself as "Reginald Herbert" (In reality the Earl of Pembroke) and motioned me to follow him. I asked the sentry if it would be OK to leave the bicycle at the gate, which it was, and then offered to leave my camera with him. "Oh, no Sir, Just don't take any pictures inside." The Duke and I had a delightful time as he showed me the Architect Earl's Bridge across the River, we toured the grounds, and afterward he showed me all through the quadrangle making up Wilton House. After a good hour, he invited me into the library in his private quarters, where a roaring fire awaited us. Tea was typical four o'clock English tea, with scones, watercress sandwiches for just the two of us. I admired the painting of the First Earl of Pembroke which hung above the fireplace. It was particularly interesting in that it looked exactly like my father. (Gale saw the same similarity on her post war trip through Wilton.) The old gentleman and I had a delightful afternoon chat which he enjoyed also, for I got the impression he was rather lonely, and did not have much company. This was probably true since he was host to the top brass of the British Army, and confined to his private apartment most of the time. One other observation bears mentioning. In the warmth of the great fire it soon became obvious that the Earl of Pembroke was not a great advocate of bathing. He probably had the same impression of me, for the climate and restrictions of heating at that time were not conducive to excessive cleanliness.

As the sun was beginning to get low in the afternoon I headed back to camp on my bicycle. The English twilights last until eleven o'clock. I should mention that of all the countries that I have visited, England's spring is the most lovely I have ever experienced. Imagine fields of blue or yellow crocus in full bloom, or clumps of daffodils in the most unexpected places. Imagine the Hawthorne perfuming the air of a country lane, while in the thicket a cuckoo livens the area. England in the spring is the most beautiful place to be, but as we told one of our friends who had just awakened from a nap. "So sorry you didn't go with us. You slept through spring and it's raining again."

Our camp site was a wonderful place from which to watch the English Spitfires swoop down on the "Buzz Bombs." The pilot would get above the incoming bomb, dive down, put his wing under the stubby wing of the bomb and tip it into the ground, causing a tremendously satisfying explosion on the South Down.

In the little village of Nether Wallop was a home belonging to a rather wealthy Englishman. He even had a tennis court (grass of course), enclosed by wire. I soon made his acquaintance, and enjoyed learning to play on the slower turf. We even liberated some sugar from the mess, and since the strawberries were in full fruit we made some ice cream from the forbidden milk. The family enjoyed it even more than we did. In fact we became such close acquaintances, that a few of us were invited to the daughter's wedding (even though I had beaten the bride and groom on the tennis court rather thoroughly).

On the day of the wedding we appeared in full dress, and all the other guests were also. (There were about fifty in all) The wedding was held outdoors in a lovely garden. A large tent had been set up for the dinner to follow later in the evening. I tasted the delicious drink in a crystal goblet (I know now) and innocently remarked to the young lady nearby, saying "This is the most delicious cider I have ever tasted." It must have been a rather special Champagne judging from her look of surprise. In my innocence I didn't know the difference.

After the wedding we were seated under the marquee, with an enormous number of heavy English silver spoons, knives, and forks plus an array of wine goblets. The linens were spectacular, and quite spontaneously I made the observation to the young lady seated beside me "These are the most lovely napkins that I have ever seen." It was said purely to make conversation. She did not speak to me the rest of the evening, and only later did I find that napkins refer to the cloths used by young women during the monthly "Curse." So sorry I don't speak English as it should have been spoken. E.G. One doesn't use the term "e's a bum" in good society, and never uses "bloody" under any circumstances.

With my interest in tennis, (My father had been Captain of the tennis team at Wofford College in 1904, and I, with both my brothers, had followed as Captains.) Wimbleton was a must. With some finagling I made it to the matches. The finals were held in the Center Court with great ceremony. Four o'clock came in the middle of the Championship Match. In the middle of a set when a game ended, the Queen rose and tea was enjoyed before the match was resumed. Four o'clock tea is still hallowed and all over the countryside it's a common sight to see lorry drivers pull over to the "Lay by" and " brew up a cuppa."

Of course London was calling as a "must see" and being granted a short leave, two of us, Captain Yoslow and I caught a ride to Winchester, then boarded a train to the big city. We had been directed to a small British Hotel near the Red Cross Club. We were late getting into London, and after settling our belongings, and washing our faces, we set out in pitch darkness to enjoy the sights of the big city. The Red Cross Club being close by, we felt sure we should start there. As we left the door of the hotel, we heard the unmistakable sound of a buzz bomb approaching. As we had been instructed, we fell on the marble floor of the porch behind a pillar. The bomb went off several hundred yards away, but close enough to shake us. We agreed "This is no place for two Southern boys" and made plans to leave London by the first train. Doctor Yoslow had received a small cut on his chin when the bomb hit, but we were not sure he hadn't hit his chin ducking for cover, so no application for a "Purple Heart" was forthcoming. The next morning we were served cold kidney pie for breakfast, and since memories of my pathology still survived, I could pick out the tubules and glomeruli of the organs served, (we agreed again this was no place for two scared Southern boys). Perhaps that episode accounts for my somewhat apathetic enjoyment of the Big City on subsequent visits. Even today I try to avoid large cities. As a rule the people are rude, the living expensive, meals are erratic, bathrooms are hard to find and always foul smelling, and after seeing the required sights I tend to get out of town.

On one of the visits to London, we saw the changing of the Guard at Buckingham Palace. It was a feeble show compared to the display of today. Only a Bobby was required to guard the palace. I never thought that I would be coming out that very gate in years to come after having had tea with the Queen in her garden and exiting through her palace. (That is another story for telling in the future.)

On a trip to London one of the nurses noticed a sign in a green grocer advertising South Carolina peaches for one pound six pence. She assumed that the price was for the package of peaches, whereas the price was per peach. This made her purchase well over $50, for she was too embarrassed to admit her ignorance. (They were beautiful peaches.)

Above the hospital site was a rather steep hill and on top was a "Long Barrow" which long ago had been plundered of the prehistoric relics of some long buried Celtic Chief. But it was a worthwhile climb to enjoy England at its best, if only for a few moments. Oh yes, a "Hill Fort" of prehistoric man was there also.

According to the local guide books, the site of Romsey Abbey was within easy cycling distance, and since it is one of the few buildings predating the Norman Conquest (1066 - one of the dates I recall), and being "Old Ruins" I was determined to visit the church. I chose a week day, thinking the Abbey would be deserted. As I neared the ancient building, the bells rang out and continued to ring on and on. This was my first introduction to English bell ringings. The sounds reverberate through the countryside. I still am not sure I understand the difference between ringing a Peal of bells, or ringing a Change of bells. I think Peals are rung on Sunday (such as Evensong) and Changes are rung during the week days. At any rate they were ringing Changes. Only one old verger was in the building downstairs, and he showed me over the building expounding the difference between Roman arches (curved) and Gothic arches (pointed). Finally he took me to the bell tower. There I met the enthusiastic bell ringers. They explained the differences, uses of the bells, their care and feeding, and finally asked me if I would like to participate in the ringing of the Changes. I was delighted and grasped the rope which disappeared through a three inch hole in the ceiling to the tower above. "Just one more thing, sir." was a final admonition. "Don't forget to let go of the rope. There's a two ton bell on the other end of the rope, and it will pull you right through that little hole in the ceiling if you don't remember!" It was really a fun morning. And to those of you who have never experienced listening to the bells of evensong on a lovely spring evening, as the clouds go scudding by and the whole countryside lies awash with the incredible sound of the "pealing of the bells," you have missed one of life's most exquisite emotional experiences.

I neglected to mention that a shared package of cigarettes, or a shared candy bar was a sure way to open many a door and made many new friends for me. This is the only time in my life that I truly enjoyed smoking a "coffin nail." Oh Yes, we got our laundry done every week for a pack of Camels. It was better than pounds or franks. This worked well with learning to ring Changes. I have not yet had opportunity to use the knowledge that I acquired that day. Perhaps someday someone somewhere will need my expertise at ringing Changes!

I recall two tasks allotted to me during this pre-invasion days. There was an American air field not too far from the hospital, out of which came the B29 bombers. One of the planes buzzed a bunch of nurses who were down on the seacoast, in swimming suits. The plane came in too low and decapitated one of the nurses.

The fly boys gave some unbelievably wild parties, so I was not at all enthusiastic about being appointed to take 98 nurses to an upcoming air force "shindig." When the trucks arrived at the air field, I made sure that every nurse knew well that the trucks would leave the base for the hospital at one o'clock sharp. When we entered the building, I was glad I had made it plain, for there was a contest beginning. Each table was set under a window, and the winners were determined by the table occupants who could fill the window sill with empty champagne or whiskey bottles the fastest. Where they got all the alcoholic beverages I never knew. (The air force had ways of getting the best of everything, thus arousing the jealousy of the other service branches, no end.) At one o'clock the trucks left the air base with only 96 nurses aboard. I still don't know where the other two girls are.

I never found out the reason for the other chore I was given until I was viewing some WWII videos recently. I was given 100 men to be trucked to Southampton. There we were placed on flat bottom, an assault craft, and taken out into the English Channel. Over the next two days we plowed up and down the Channel. These craft were not planned for pleasure boating, and inevitably many of the men got seasick. Then we were taken back to shore, at Weymouth, and returned to the Hospital at Winchester. Nothing more was said about the incident, and I learned only later that this was done to find out how long a unit could be kept in the Channel until rendered ineffective by seasickness. Our group was worthless in 48 hours, having over 50% leaning over the side.

About this time I was sent on detached service to a British Surgical Hospital, to learn chest surgery. It was a busy service, but about all I recall was treating two tough British commandos with similar wounds. It seems that one of the commandos would not believe the blanks used in the war games would cause any trouble and proceeded to put his gun against his buddy's chest. When he pulled the trigger the blast and the wadding blew a hole in his chest. In the barracks that night the story was not believed. "Gor blymee! You mean jist lak this?" And another incredulous commando blew a hole in the teller of the tale's chest. The two casualties sustained almost identical wounds.

Shortly before D-Day (6 June 1944) some idiot in the upper echelons decided that all medical officers and nurses should be qualified in the use of the .45 caliber pistol and the carbine. It was my job to take them to the range. The target for the hand gun was fifteen feet away with a five inch bulls eye, on a five foot target. Most of the participants plowed up the ground in front of the target, while the target keepers of the carbine range couldn't keep the red flags going fast enough displaying "Maggies Drawers," which indicated that the one who had fired had missed the target entirely. It was a complete fiasco. All I got out of it was ringing ears for being on the range all day.

We in the 48th General Hospital knew when D-Day was on before the rest of the world did, for we were treating the casualties of the Invasion Air Fleet during the night of D-Day. When a plane was hit by flak, and the parachutist was injured he was brought back to England. Therefore we operated all night, and were informed by the patients themselves that these were the invasion forces landing in France. *The Medical Services in the European Theatre of Operations,* states, "Another heavily used unit, the 48th General Hospital at Stockbridge, processed about 3,000 casualties in eight days."

Among those injured was a young paratrooper who had been shot by a piece of flak while in the air. The X-ray revealed that the metal was lodged in the thick muscle of his heart. At first he seemed in good shape, however soon after admission he began to develop signs of heart failure. Another x-ray showed that the pericardial sack surrounding the heart was filling with fluid. With a quick insertion of a long needle I relieved the congestion. But this obviously was only a temporary solution. I called the colonel and set up the OR for an immediate thoracotomy (exploration of the chest). As I recall we called for blood but did not wait for a cross match. The chest was opened and there was the offending body bobbing up and down with each beat of the heart.

I carefully inserted a large needle, with a heavy suture swedged on, around the metal fragment, taking care to avoid jeopardizing the coronary artery, and surrounding the fragment of flak with a purse string type of suture. The needle insertions had to be made during the asystolic phase of the heartbeat when the heart was quiescent. Remember, this was years before heart machines allowed surgeons to operate on a heart which was chilled and still. The instant the metal fragment was removed, there was a gush of blood quickly controlled by ligating the previously inserted suture. There was no further bleeding and the closure was routine. That was one lucky boy, for if the fragment had penetrated another quarter of an inch, there would have been a large hole in the large chamber of the heart, resulting in instant death.

A short while after the initial landings we were relieved of duty and another hospital took our place. We were sent to the seashore rest area. Again I was selected to go as advance unit to prepare the site for our embarkation. The squad arrived at these tremendous cowpens, and were informed that our job was to clean up a hillside filled with "honey buckets." For you uninformed, this title referred to the buckets inserted under the latrines. These had to be cleaned and sterilized with high pressure steam hoses after the departure of the previous unit. At that time, we experienced the nitty gritty of soldiering, i.e. cleaning up a real mess.

Some 24 hours later our group arrived. However we were not penned in, but went straight through to the Star of India, a decrepit liner of the Indian ocean which took us across the Channel that night. The next morning the beach at Red Sugar Utah was not ready to receive us and the English and Indian sailors had to eat the same rations we received, i.e. K Rations. They griped for they said they had steaks in the freezer, but could not eat them until we disembarked. We wasted little sympathy, for we were landing on French soil to free the world from Evil. We crossed the English Channel on the 6th of August, some two months after D-Day. We believe we left from Falmouth. Doctor Fredrickson recalls talking to several sharpshooters who were posted on the prow of the ship. These were to fire upon the bloated bodies of dead GIs which occasionally rose from sunken ships, and the sight of which might have upset oncoming replacements.

Climbing over the edge was no fun, and I'm told that the back packs we bore weighted eighty pounds. The packs weighted much more by the time we struggled over the long flat beach. We were happy to see our bedding rolls being loaded onto the trucks as we trudged along with Mere St. Église in the far distance. That night was spent in the field near that famous little town. I'm sure we had pup tents but I don't recall putting them up. Again Doctor Fredrickson recalls that at Mere St. Église we camped across the road from a gasoline dump containing millions of cans of gasoline.

The following day we were put in a field that had been de-mined and then we were forgotten. The field was about the size of an American football field, and in one corner was a little hut which

Gale and Bill III - 1944

My PINUP
Gale - 1942
The "Infallible Critic."

The Tullahoma Family

England - The old Gates of Southampton.

Headquarters, 48th General Hospital, Stockbridge, England. Fire Reservoir - no mosquitoes in England.

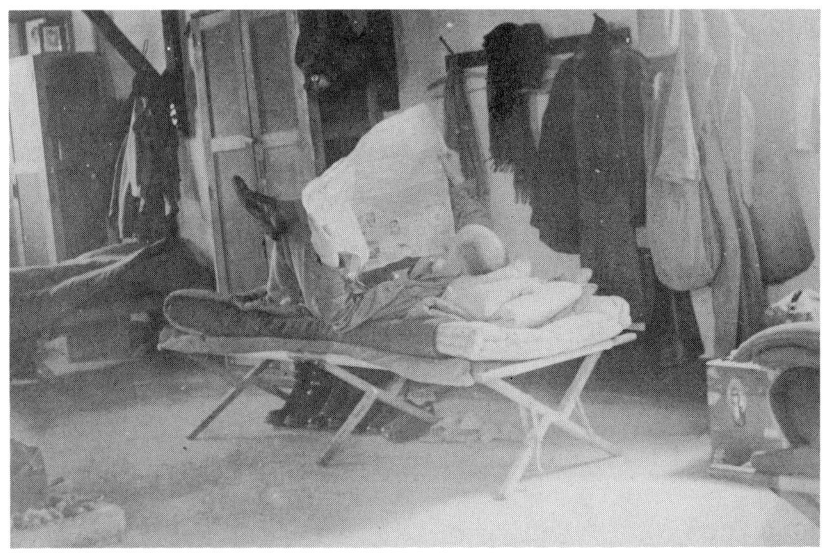

Officer's Quarters, 48th General Hospital, Stockbridgek, England. Colonel Kellog improving his mind.

Buckingham Palace - 1943
W. C. H., Major "Buck" Williams and total Palace Guard.

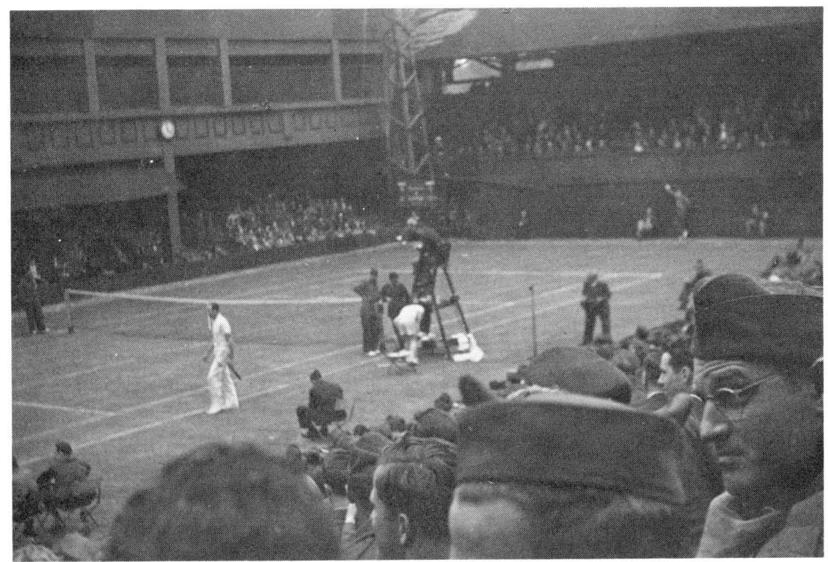

Wimbleton - 1943. The match has just been interrupted by Her Majesty rising to go out for her four o'clock tea.

Dover - 1943. The castle in the background.

The 48th General Hospital disembarks from the* Star of India *onto Red Sugar Utah Beach on the shores of France.

The 48th General Hospital crowded into a LST landing craft approaches Red Sugar Utah.

The 48th General Hospital trudges across the beach in France carrying eighty pound packs.

U. S. Army Photo

Nurses of the 13th Field Hospital
First to land in Normandy to tend the wounded,
take time out for a meal.

Paris following liberation
Note traffic congestion around the Arc de Triomphe.

Site of the Massacre in the French Villiage.

Encampment at Mère St. Église, France.

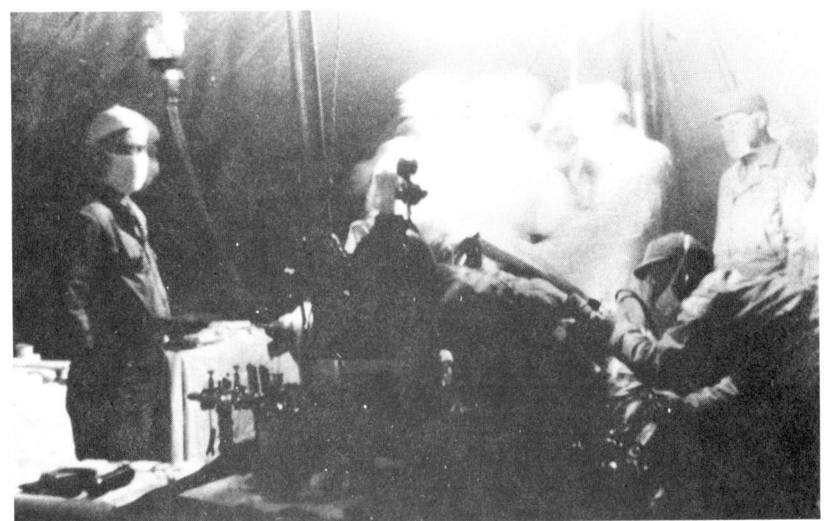

Operating Room - 54th Field Hospital
September, 1944, near Metz.

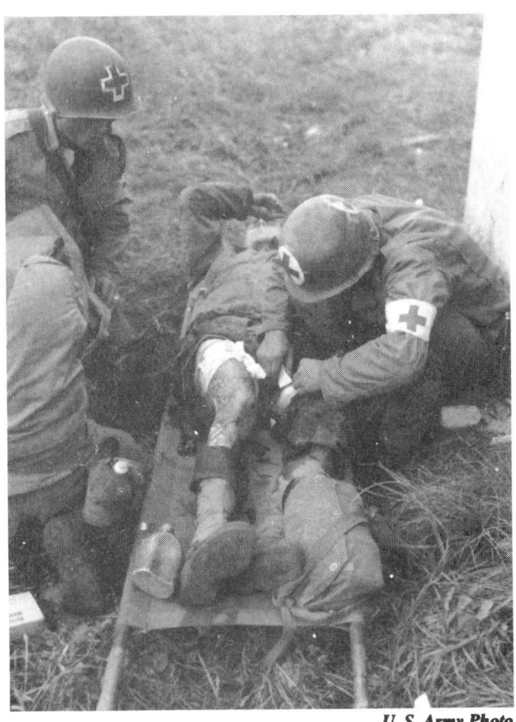

U. S. Army Photo

Medics on the Front Line

U. S. Army Photo

Battalion Aid Station personnel readying casualties for the next stage of their rearward evacuation.

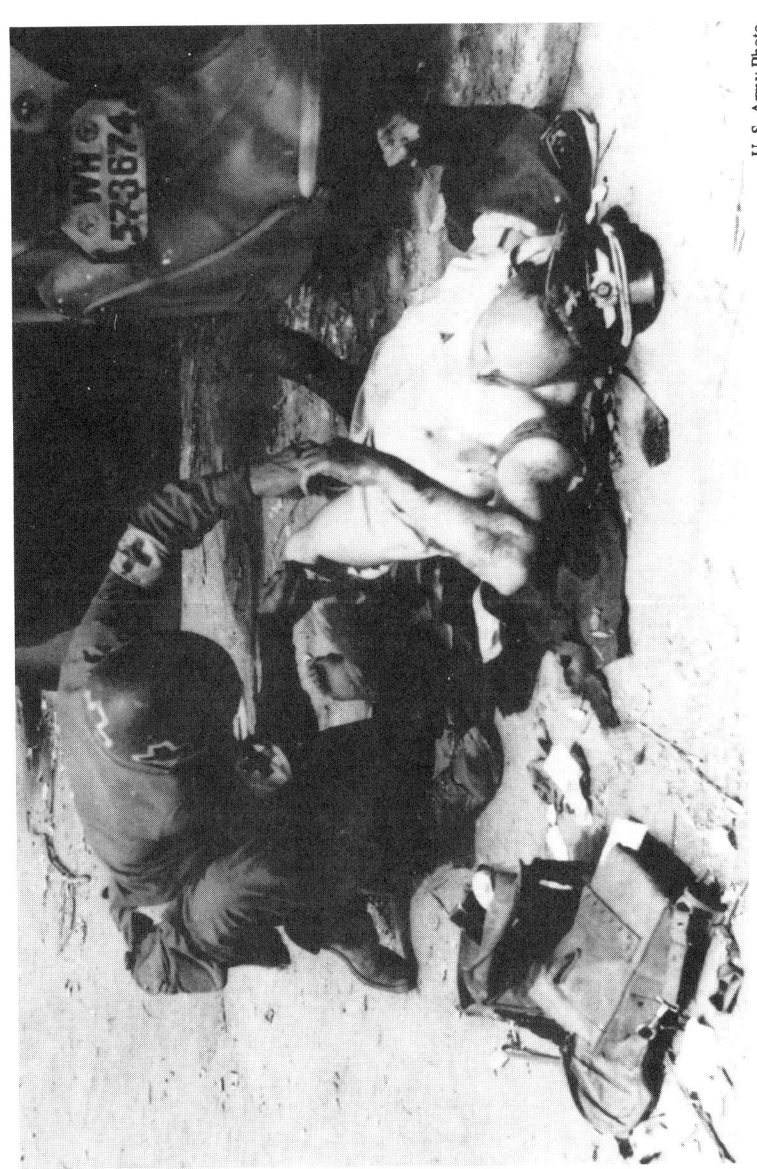

U. S. Army Photo

Medic treating a wounded German soldier,
one of the thousands swept up in the pursuit across France and Belgium.

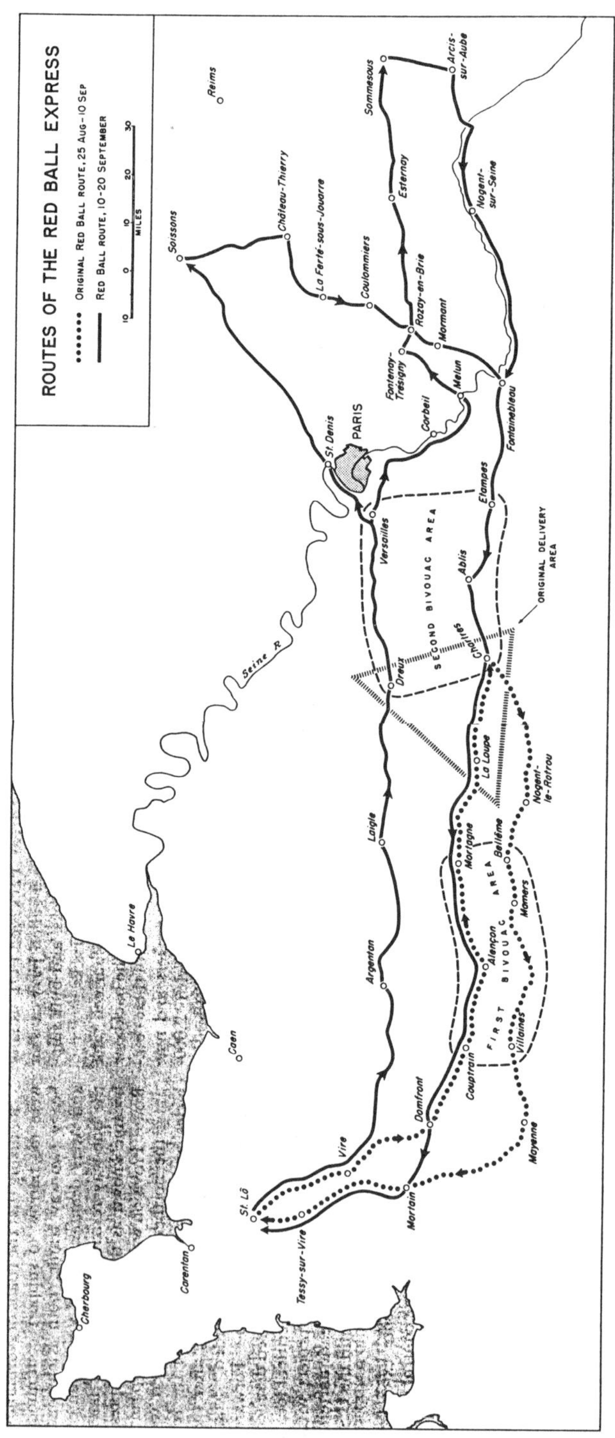
ROUTES OF THE RED BALL EXPRESS
ORIGINAL RED BALL ROUTE, 25 AUG-10 SEP
RED BALL ROUTE, 10-20 SEPTEMBER
10 0 10 20 30
MILES
Reims
Soissons
Château-Thierry
La Ferté-sous-Jouarre
Coulommiers
Sommesous
Arcis-sur-Aube
Esternoy
Nogent-sur-Seine
Rozay-en-Brie
Mormant
Fontenay-Tresigny
Melun
Fontainebleau
Corbeil
PARIS
St. Denis
Versailles
Etampes
SECOND BIVOUAC AREA
Ablis
ORIGINAL DELIVERY AREA
Seine R.
Chartres
Dreux
La Loupe
Nogent-le-Rotrou
Laigle
Mortagne
Bellême
Mamers
Alençon
FIRST BIVOUAC AREA
Argentan
Villaines
Couptrain
Domfront
Mayenne
Vire
Mortain
St. Lô
Tessy-sur-Vire
Le Havre
Caen
Carentan
Cherbourg

Gate of Hospital Lariboîsière, Rue Ambroise Paré.

Aboard "forty and eight" (forty men or eight horses). From Metz to Marseilles took four days.

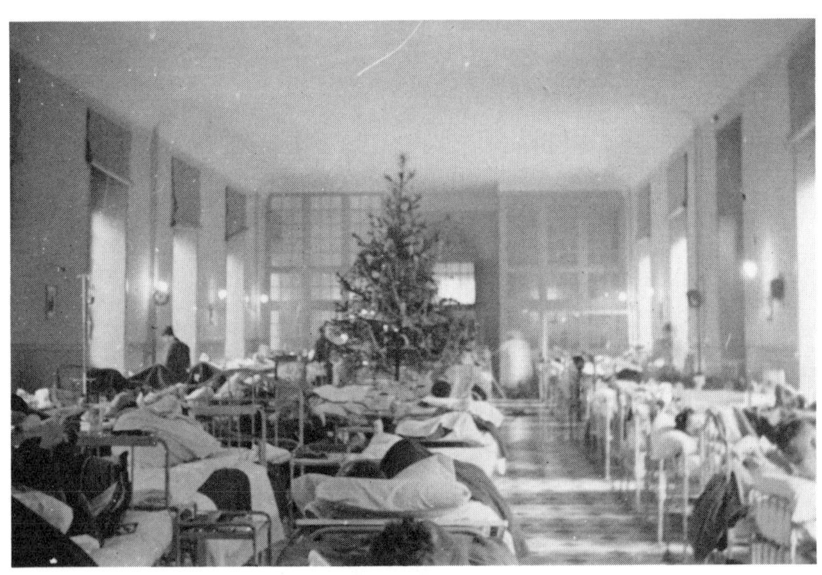

Christmas Day - 1944
"L'Hospital 48th General", Paris

smelled so bad that the GIs poured gasoline over it and set fire to it. In a short while a rabid French farmer came roaring through the gap in the hedgerow. Since I had a good deal of French in college and had even won a National Scholarship to McGill University for writing French poetry, I was appointed the unit translator. (Please note: I speak South Carolina French at two hundred syllables per minute, whereas the French speak at six hundred syllables per minute. Therefore they could understand me quite well but I hadn't the slightest idea what the Frenchman was saying. Nearly every sentence of mine was preceded by "Plus lentement, s'il vous plait! i.e. More slowly, if you please.") It finally turned out that the GIs had burned up the farmer's yearly supply of cheeses which was curing in the little hut. We parted friends. He even pulled out his knife, which he wiped on his trousers leg, before cutting me a slice of his Pont l'Eveque cheese. I hope the American Adjudicator settled his claim eventually, for it was superb cheese.

The GI's sense of humor, ever present though somewhat warped at times, was evidenced in the little field confining some three hundred of us. The officer's slit trench was properly dug with the sides straight down, but the Nurse's slit trench was lovingly tapered at the top. This may have sufficed in dry weather but France was wet most of the time when we were there. Inevitably at least once a day we heard a scream and cussing announcing that some luckless nurse had fallen into the trap.

This was the area of France which has a favorite drink called Calvados. This is a highly potent drink made from the distillation of apples. Although we were confined to the unmined field, somehow the GIs made friends with the surrounding population and experienced the delights of Calvados. Some said that it made good lighter fluid, and some even had the courage to try drinking it. I recall one soldier saying that if you drank a large tumbler full it would "make you awful sick!" It was here we stayed until the fall of St. Lo. (27 July 1944)

As usual, the army was living up to its reputation: Another SNAFU. They had no need for General Hospitals at this time in the fighting, some ten days after the D-Day landings. But they did need surgeons with the field hospitals. I was fortunate. I was assigned to take a surgical team to a Field Hospital following General Patton's

Forces. That night we were taken to Chartres. It was a moonless night, but I could see the towers of the famous Cathedral, for we were parked in the square in front of the building. I was the senior officer and also interpreter so I was given a Gulf touring map to follow the Red Ball Highway across France to join our designated Field Hospital.

If you have never been through the French countryside in a complete blackout, surrounded by the ever present boccage, (hedges) at night, in those huge army trucks, into unknown territory with an inadequate map, then you cannot know the sense of adventure it occasions. When driving in a blackout, the driver follows the truck in front, which has three small lights shining through holes in the rear bumper. If the driver sees two lights he is at the correct distance, if he sees only one he is too far away, and if three appear, prepare for a collision. All night in that fashion we lumbered to Fontainebleau where we stopped for the day. We were sent to Fontainebleau Headquarters to get our orders. To kill time, we went through the Chateau. I recall the guide emphasizing the carp ponds where the executed criminals were fed to the huge carp, and the bear pits in the center, but I was particularly impressed by the inlaid flooring and the painted ceilings. Why we ended up in Fontainebleau from Chartes I will never know, and my long time friend Captain Fredrickson was no help; nor why we spent the day in loafing around the Chateau. Mysterious were the ways of the army.

The road to the front ran right through Paris, but first we had to obtain better maps. We went to a Division map tent but the Captain in charge refused to provide us with a map for such a small unit as four trucks. My Sergeant pulled me to one side, and under his breath, he said "Keep him occupied, Major!" and he disappeared with the Sergeant in charge of maps. In a few moments he indicated it was time to leave. When I checked his acquisitions, I found that we had maps so precise that the footpaths of the region we were seeking were clearly delineated. I never inquired how the maps were procured, but I had the feeling that an amount of hospital (100%) alcohol changed hands along with the maps.

Since the direct road to our destination led through Paris, naturally, in an effort to save gasoline, we chose to drive through the City of Light. Since we had to stop for lunch, this made an ideal stopping place. The place chosen by our drivers, who had never been to Paris before was the Cemetery de Pere LaChaise. How he knew that this was quite near the Red Light District of the City was never explained, nor did I enquire too eagerly. I assumed that all of the contingent was anxious to see and learn of the treasures of this unique city. We saw, with the aid of the FFI, (Force Francais de l'Interior) who were the French underground which had liberated the city (remember that this was only a few days after the liberation of Paris). I was called "Monsieur Le Liberateur" on several occasions. The FFI (Paris was liberated 18 August 1944 by the French Forces of the Interior) bore guns of all kinds, and explanations had to be made that the Women with the group were "Les Infirmieres pour le blessé," and not "Les femmes de l'Armee." (Since all French nurses were nuns, the appearance of our nurses in army uniforms confused the French people.)

We were disappointed that we could not go to the top of the Eiffel Tower but only to the first landing. The rest had been taken over by the military, but we did get up on the top of the Arc de Triumph to find it guarded by FFI men who welcomed us all enthusiastically, hugging the nurses again and again.

To my surprise all the contingent appeared at the appointed time, at the appointed place and we drove out of Paris feeling the stop over had been highly successful. We stopped for the night near the Malmaison, former home of Napoleon and Josephine, about twenty miles outside of Paris.

The following day we joined the Red Ball highway across France from the invasion ports to the front. We only got off the correct road one time by making a wrong turn to the right on a narrow country road. We were beginning to feel something was wrong when a burst of machine gun fire hit the kitchen truck making such a noise among the pots and pans that the drivers immediately turned around. If you can't imagine any of those large army trucks turning around on a narrow country lane, then you can't conceive the impetus that machine gun fire can inspire. It turned out that the fire was from American airplanes. The panels on top of the trucks had not been changed to conform with the day's new orders. Fortunately the second plane did not fire after seeing us turn around.

We came upon a German supply dump which the tanks had plowed through not long before. Among the other supplies were medical and surgical equipment. Since the Continental surgeons were much more quick to do an amputation than we were, one of the indispensable tools of the trade was a long and very sharp knife called a bistoury. The following morning, after passing through the supply area, all of our enlisted men wore bistouries. They quickly disappeared when it was explained to them that this was strictly against the Geneva Convention, and the Germans would shoot them on sight if they were wearing such a weapon.

Around about these parts we passed the tiny town of Dom Re Me, which name stuck in my memory as the birthplace and childhood town of Joan of Arc. Since we only saw her backyard and passed by rapidly, I hardly felt like asking for her blessing on our endeavor.

We occasionally saw piles of dead Germans. The thing most often noted was that none of the bodies had on shoes or boots. The French peasants stripped the bodies of all footwear. When the GIs were brought in wounded we removed the footwear outside the tents, since it would not be needed during the hospital stay. These boots were left outside admissions and triage, and were rapidly removed by the French people.

Traveling in convoy on the Red Ball Highway across France was a long and tedious affair. Stops were few, for getting back into line again wasn't easy. The question of the men voiding apparently wasn't much of a problem, but for the nurses it was a different matter. They solved the problem by erecting a blanket in the back of the truck and the steel helmet made an acceptable though unstable receptacle which was emptied overboard into the French dust.

At this time we joined the 54th Field Hospital which was attached to the 78th Division, following Patton on the mad race across France almost to the German border. Most of us had little use for the Commanding general. We felt that he carried the "pearl handled" revolvers to keep some of us from shooting him in the back. I felt and still feel he was a glory seeking SOB, caring only to have his name in the history books. We knew from personal experience just how close the Germans were on our flanks, and the incidence of his allowing the tanks to go until they ran out of gas was still fresh in our minds because we were close behind. The Germans clobbered our tank men whose turrets couldn't even move because of running out of fuel. I still feel that any General who doesn't know his logistic support, or care should be court-martialed.

A field hospital of WWII was very similar to the M.A.S.H. unit of TV fame. It consisted of two or more sets of tents just behind the front lines. When the front moved on, the support team moved forward, set up the tents, and had everything ready for the surgical team to follow. The Surgeons finished the Operating Schedule and left the wounded in the care of the Medical officer and Nurse who, using a skeletal crew, evacuated the recuperating wounded to a General Hospital. The ORs set up were right behind the front. I had a soldier on the table for abdominal operation only twenty minutes after he had been shot. The artillery frequently pulled in behind us and the missiles could be heard as they flew over our heads. This could be unnerving, but even that came to be accepted. I don't think that was exactly "cricket" however. I frequently waved to the pilot of the German Storch (a small slow flying plane used as an artillery spotter) as he cruised over the hospital looking for the artillery situated behind us.

At one time, Gale knew almost exactly where we were, for Bing Crosby reported the he had visited a field hospital near a French town I had mentioned in a letter. This we all recall as the most emotional episode during the campaign. In the Recovery Tent was a young man who had received numerous wounds and was obviously dying. He was a favorite of all of us. When he requested that Bing sing "White Christmas" it set us all to weeping. He died the next day. Alone I performed a complete autopsy that day, in the open field behind the hospital. I can still recall the lovely autumn sky and the end of summer feeling. All during the autopsy I wept unashamedly throughout the procedure.

The OR may have seemed a little primitive but the surgery was not. It and the instruments and the anaesthesia were on a par with surgery anywhere. The lights left something to be desired, for they were constructed of large cans which had held canned fruit. The table was a litter gently lifted onto two saw horses. I pause to pay tribute to the medical corpsmen. In all circumstances they were always gentle. And they would work until they dropped. In fact all of us did until it was found that if a surgical team worked beyond eighteen hours at a stretch, the recuperation time made it inefficient. Therefore the CO had to call a halt at times and order a rest. The floor was a canvas tarp, which quickly was ground into the mud. On the center pole of the OR tent was a sign admonishing us "Don't Throw the Kidneys and Spleens on the floor! It makes it slippery." I operated in undershirt, trousers, and field boots, with the sterile gown thrown over all.

One morning there was a heavy rainstorm in the hills above us and unfortunately the water channeled in our direction. In a short while the flood came through and we were soon operating in ankle deep water.

The operating table was a stretcher set on two wooden sawhorses of equal height. If Trendellenburg position was desired, (i.e. the head of the patient needed to be raised) a sawhorse of a greater height was inserted.

During one difficult procedure, Doctor Fredrickson recalls the the patient wasn't doing as well as expected. He states that I leaned over the top sheet to talk to Doctor Bickel, the anaesthetist. Doctor Bickel had his head between his knees. "What are you doing?" was the question. Doctor Bickel replied in a muffled voice "I'm praying." I'm sure we all were praying at times without number.

Doctor Bickel was a Polish Jew who escaped after Hitler occupied Poland. His field was psychiatry. However he was also a very excellent anesthesiologist. He died not very long ago. He was supposed to attend the 48th General Reunion at Fort Lauderdale two years ago, but called to let us know that he was ill. Doctor Lynn Fredrickson and I seem to be the only two remaining physicians of the unit.

One of the patients was a Nazi SS Captain who insisted on being first in everything, and complained to everyone because he was in the ward with enlisted men. I'm glad I did not have the duty of triage to determine the order of operation. He was operated in his turn however, but he still stayed in the room with all the rest. In fact, we made no distinction in the rank of the wounded.

One episode I regret comes to mind. A French Cure brought in a boy about twelve years of age who had dislocated his elbow several days previously. We were asked to reduce it, and not having the necessary orthopedic skill, we found that an Orthopedic surgeon was in the next echelon behind us, and the lad was sent there, I hope. I should have made an attempt to reduce the injury.

It was getting on late in summer, and none of us had had a bath since we had left England. Looking back I cannot recall that any personal body odors were ever noticed. We received a helmet full of hot water each mornin. With that I shaved first, washed my face, then the rest of the body parts. Using the remainder of the liquid, I washed my underwear, and finally, if any fluid remained the socks were cleansed. Thus when the call came late in the month of August that the shower truck was down by the river, we were all anxious to go. Fortunately there was a lull in the fighting at the time.

The shower consisted of a curtain wall about fifty feet square with numerous shower heads and wooden floors. The nurses were allowed to go first, of course. But they stayed and they stayed and they stayed. So far as we could tell no attempt was made by the shower crew to get them to hurry along. Investigating, I found the shower boys were making a fortune renting out a peep hole to all comers. They would still be there had I not gotten the attention of one of the nurses. When she came outside, she was not informed of the peephole, but I did take her down to the river and show her where the intake pipe was getting the shower water. A long dead cow, greatly bloated, had drifted down the lazy stream and was circling around the intake valve. She screamed and ran back up the slope to tell her friends. They quickly left the shower square. At which time we pushed the cow down the stream, and enjoyed the first bath since June 16th. Wonder how that cow happened to settle over the intake valve?

Living in a pup tent could be hazardous. Early in this epistle I mentioned that Gale and I had bought inflatable mattresses. I was perspicacious enough to have brought one along with me in my bedding roll. Now a pup tent doesn't have much head room, so since the weather had been rainless for a month, I pitched my tent across a slight gulley on a hillside. During the night, miles away there was a rain, and that night I awoke to find myself bumping gently against the head of the pup tent. The water had flowed down the ditch to float my air mattress without waking me. I still felt that buying the air mattress was one of the best purchases I ever made. One of my buddies in France offered me $100 for it, and I refused.

One of the most emotionally stirring episodes of the war was experienced late one evening. Lying in the ditch, in the pup tent, I was startled to hear airplane engines overhead. Then the British Broadcasting music came on and the sounds of Beethoven's Fifth Symphony filled the night air. We later learned that there had been a thousand plane assault on a major German target that night. It was well over two hours before the sounds of the attacking aircraft faded in the direction of German territory.

We were lucky to have a radio. Some of us had urged that the Supply Sergeant try using the 98% hospital alcohol as an inducement to acquiring one. And indeed a magnificent German Blaupunkt soon appeared. We enjoyed its sounds and wide selection until rumors that the General's jeep was missing a radio. Our radio was replaced by a muchly inferior model and the Blaupunkt model disappeared. Of course the General's radio had been liberated from some German command site. Depend on the Supply Sergeant to acquire anything really needed.

During one of our advances toward the front, we came to a little French village in deep mourning. It seems that the day before, the village had heard that the Americans were coming and the population lined the one village street to wave flags, throw flowers and give a generous welcome to the liberators. Only it was not the Americans who appeared, but the retreating Germans. The Huns separated the boys and men, and shot them all. The village was desolate. We stopped our journey long enough to attend the mass funeral in the tiny old church.

The following day we stopped at a little village and the people gathered quickly. In our trading for fruit, vegetables and memorabilia, I arranged a swap of a carton of cigarettes for a Voightlander portrait camera, which is supposedly the tops in such instruments. The Frenchman went off to fetch the instrument, and in the meantime our column of trucks pulled out. So go the fortunes of War.

I have mentioned my admiration for the Army procurement officers. While we were up front, our officer appeared with a truck full of enormous black bearskin shakos (these tall hats made the crack drill teams look larger than real life) taken from a German supply dump which had been overrun. We all acquired a black dress shako for which we had absolutely no use, for we never wore such military adornment at our parades. But our ingenious supply officer eventually swapped the truckload of dress hats for a truckload of champagne, Cointreau, Benedictine, and Cognac, which was quite useful.

Up near the German border the fighting stalled and many of us thought that Hitler would soon see the handwriting on the wall and surrender. Little did we know the extent of his madness. Since winter was approaching, we were glad to get a call to reappear in our old unit, the 48th General Hospital, in Paris of all places. The 48th General Hospital began to admit patients on 23 September, even though there was the usual army SNAFU. From *Medical Services in the European Theater of Operations* came the interesting note, "Inevitably units became separated from their equipment assemblies; The 48th General Hospital went to Paris while its outfit went to Liege."

How our Colonel Alling wrangled this choice assignment we never knew. He was, of course, regular army and had been an excellent Orthopedic Surgeon. He had written a book on the subject. He made the statement early in the forming of the hospital that he knew good surgery when he saw it and would not interfere in the medical practice if he could avoid it. He never did! But how he achieved the assignment to Paris remains a mystery.

We took over The Hospital Lariboisiere, a two thousand bed facility close by the Gare de Nord Railroad Station. It had been a French General Hospital before it had been taken over by the Germans. When the Germans took over they cleaned out the wards and shoved the debris under the wards. When the Americans took over they shoved the German equipment in the cellars below the wards. A century and more before our time, the hospital had been donated to the city by the Countess Lariboisiere in gratitude to the Lord since she and her two children had escaped the dreaded plague which was raging at the time. The hospital was built in 1835. A statue of the Countess and her two children stood in the center court.

At the distal end of the courtyard stood a beautiful little chapel, Catholic, of course, in which services for all faiths were held. One morning it was announced that only Catholic services would be allowed in the chapel. Colonel Alley was incensed. One of the Freedoms for which we were fighting was Freedom of Religion. He made an appeal to the big wigs of the Catholic Church in Paris and received a flat "No!". He put a padlock on the door of the Chapel saying nobody could use the building. The fat was in the fire! Nobody, but nobody, treats the Catholic Church in that fashion. The edicts went all the way to Eisenhower's Headquarters before it was resolved. "The Chapel stays open for use of all faiths." And so it remained, at least as long as the Americans held the Hospital. I think Colonel Alley deserved the highest medal for sticking to his guns.

The "Three Mousqueteers" arrived from the front lines, all three equipped with face herbage. Captain Fredrickson had a goodly covering and Captain de Lucca maintained a magnificent hirsute adornment. I, however, barely managed to sport a tiny moustache and a small pointed chin covering. I thought I looked like Ronald Coleman, but Mephistopheles was more probable. The facial growth did not survive long, however, for the Colonel let it be known the army regulations of some kind frowned upon such adornments. I resumed my duties as Chief of General Surgery. The Triage Center, the Chest Center, the G.U. Center and the Neurological Center for the whole area were located there in the 48th General Hospital.

One of the Nurses wanted to get married in the Chapel and so it was agreed. However she wanted me to sing a solo during the wedding. I demurred because I had no music or the words, but she was smarter than I. She had written home for both. I sang, she got married as a Protestant, and the Chapel did not fall down. I'm sure it was properly fumigated after the "heretics" left the area.

We encountered one problem I've not seen before or since. The Russians were allies, of a sort, and the Russian Military Group was attached to Headquarters in Paris. One of the young officers developed acute appendicitis, and was sent over for our care.

He did have acute appendicitis and I removed the offending organ without any difficulty. The following day the Russian General and entourage arrived to check. We, for the first time, had made an exception and placed the young patient in a private alcove which was glass enclosed at the front. Everybody was very happy, in spite of the fact that the young man spoke no English and my Russian was "Nyet." The patient progressed nicely until we judged that he was able to return to duty. Then multiple symptoms began to appear. After some time, it became apparent that he really did not want to return to the Russian Compound. The diplomats worked that one out and we were not involved.

One night we received a phone call to get everything ready. There was a train in the Station with five hundred casualties from the Battle of the Bulge. (December through 20 January 1945) I really can't say how long we worked that time. It was a long, long time between sleeping. The Nurses and ward men were magnificent.

The fighting was sporadic but flow of patients was continuous. On 25 March 1945 Hitler made an unexpected and unintentioned gift of the intact Remagin Bridge. This portended a rapid end to the war in Europe and resulted in the largest encirclement of enemy troops since Hannibal encircled the Romans at Cannae (216 BC).

The closest I ever saw a General come to getting kicked in the bottom occurred at that time. Of course an inspection officer had to show up about then. He inspected the ward and grunted, and then went to the room where the food was prepared for distribution to the patients. After making the usual inspection, he took out his pocket knife and kneeling on the immaculate tile floor he ran the knife down the length of the interval between the tile, and triumphantly showed us all a fragment of dirt on the knife. I would have endorsed the action if the Sergeant had kicked the exposed bottom, for we had all been working like slaves. Fortunately he held his peace, but his face expressed the way we all felt.

One of our other inspectors was Doctor DeBakey, the famous heart surgeon. He spent quite a while in the hospital. I wish I could read his report. He had an official title I'm sure but you wouldn't have known it, for he turned up in a wrinkled civilian suit, and was not pretentious at all. Quite a guy.

Another inspector was General Morgan from Vanderbilt. He was dressed in the finest tailored uniform, pressed to perfection. He was accompanied by two aides. (Doctor Debakey had none.) The contrast between the two was so obvious that I did not bother to disclose the fact that I was a Vanderbilt graduate and previous member of the house staff.

One of the wards was devoted to the care of the GIs who had acquired gonorrhea. One of the jokes current at the time was the huge royal palace taken over by the Public Relations Office had a big sign across the front of the building "PRO" which to all the soldiers meant only one thing, i.e. Prophylactic Office where everybody exposed to venereal disease was to report immediately following a sexual encounter. It kept the Public Relations people busy for weeks explaining the difference.

With the magic of penicillin, the infected troops were soon able to report for duty. This did not suit a great many of them. They would be well and immediately come down with an acute recurrence. We knew that the girls were not entering the hospital so I took a troop down under to see if the basements communicated with the streets. They did. Under the Wards was a warren of tunnels. I accomplished nothing toward blocking the tunnels, but I found some marvelous antique instruments. A monaural stethoscope of the era of Lannec (1822), a German patient to patient direct transfusion set, a set of aneurysm needles, and other "goodies" too heavy to carry back in my bedding roll. I wonder if the treasure trove is still under there? Incidentally, all the urine specimens had to be collected from the patients receiving penicillin. These were given to the French and the antibiotic recovered for reuse. The vial of 200,000 units of Penicillin sold for $200 on the black market. One Medical officer was court-martialed for such dealings.

One of the evening entertainments, when we couldn't leave the hospital area, by reason of being on call, was to watch the PRO area in the back of the Hospital. PRO stands for Prophylactic to the average GI. (Not Public Relations Office.) This involved the vigorous use of soap and the injection of antibiotic solution into the penile urethra. The French "Filles de joie," ladies of the evening, would bring the happy GI to the PRO station, and watch the procedure to make sure that it was properly performed, before she took him by the arm and departed for an evening of fun and games.

The "Pissoirs" of Paris deserve a mention. These are unique to the French, I hope. The areas are self-advertising, for if one is within fifty feet of the unit, the unmistakable ammoniacal odor of uncared for male urine pervades the region. The kiosk consists of a band of metal about ten feet long set on metal posts,so that it is elevated two feet above the ground. The feet and legs of the occupant are well visible. The covering area extends to mid chest. This allows the visitor to continue his ongoing conversation with his companion, be they male or female. The back of the booth is also metal but extends from head height to the ground, ending in a trough which may or may not have a trickle of water desolately going from one end to the other, and accomplishing little. I have used these facilities when the situation reached desperate proportions. I would survey the surroundings for possible female occupants, having found from previous experience that when these creatures appeared previous conditioning prevented the organ from functioning properly. As soon as possible after partaking of the comfort of the Comfort Station, one left the booth, almost overcome by the situation and the fumes. Occasionally, in the hurry of an exit, one might forget to zip up.

The Abbey St. Denis, on the outskirts of Paris, and the first construction to use the new fangled Gothic Arch (1166 A.D.), had a twist for this answer for the human male needs. Along the right hand side of the wall surrounding the Cathedral, are a dozen smooth rock slabs ending in a trough carrying off the water. These "Pissoirs," for such they are, have no protective shields, and were used during the Middle Ages just as the more modern ones are today.

Evidently the French females have developed a tremendous bladder capacity, for the kiosks for the girls are seldom to be found. When they do appear in some hidden corner, they consist of a large porcelain plate, a good yard square, which has two elevated footprints one on each side. There is also a six inch hole toward the back. This should be sufficient to take care of the needs, but it is a trap for the uninitiated. At completely unannounced periods, a large tank full of water hidden above, lets go in order to flush the whole area. Woe betide the unwary occupant. It is much better for the female to search out the nearby Metro Station and pay the steely eyed Madame the few franks for the paper than to take a chance of an unwanted bath.

One of the perks of the best rank in the army, the Major, was the privilege of going to the Hotel Georges Cinq. There only the field grade officers were allowed to participate in the pleasures of the palate produced by the best chefs the armed forces could commandeer. It was unbelievable what a French Chef could do with C Rations and K Rations. Even Spam attained gastronomic elegance under the ministrations of culinary magicians. (But back to the subject of this editorial.) The rest rooms of the George Cinq were gleaming white, and well worth the trip, if only for the purpose of hand washing. Here too, the panels of the "pissoir" were gleaming white, and a thin cascade of perfumed water fell into the appropriate trough. However one unanticipated hazard presented itself. An officer could be in the appropriate stance and comfortably along with his appointed chore when the lyric notes of "Bon Soir, Monsieur" coming from an unmistakable lovely female voice behind him, caused his unsuspecting sphincters to go into complete spasm. Upon looking around, he perceived an angelic apparition of a French Maid, clad in a perky lace cap, a black off the shoulder uniform, which swirled out from the waist, and ended in long tapering legs and high heels. After completing her task of delivering a pile of fresh white towels to the appointed spot, she smiled sweetly to the assembled officers murmuring "Au revoir, Messieurs," and withdrew discretely, allowing them to complete their appointed tasks, whatever that might be.

Among the patients I recall was an American tank gunner who had a bad burn on the back of his right hand. At this time we were all incensed at the Malmedy Massacre, where the German SS unit of tankers had captured around a hundred of the American soldiers and then machine gunned them. He told me the realities of war. "We got no way to keep prisoners, Doc. We take their watches and money and tell them to run, and when they do, we shoot them." I elevated a flap of skin on his abdomen, sewed the flap to the back of his hand and put him in an upper body cast. He was allowed to go into the city frequently. I made the remark that I had not been able to acquire a Luger pistol in my travels. He offered to get me one. During one of his trips outside the compound, he brought back two pistols, one a Chanticleer .25 caliber and the other a Berretta .32. He was apologetic that he couldn't get a Luger in Paris. The hand did well and the pedicle was severed leaving the back of the hand well padded.

In contrast to the European surgeons, we expended much effort to keep from losing a patient's limb. One of our troopers had an injury to the blood supply of the left arm and hand, which responded beautifully to a block with novocaine to the stellate ganglion in the neck. After several blocks, it was decided by the new Colonel of Surgery from one of the Big Medical Schools that the removal of the stellate ganglion in the neck would give the best results. The operation was set up for the following morning, with myself as assistant. The time arrived but the Colonel appeared in dress uniform.

"Have you done one of these before?" was his first question. "No Sir." was my reply. "Did you read up on it?" "Yes Sir." "Well go ahead and do it. I have to attend a meeting this morning."

The surgery went beautifully, and the patient got an excellent response.

Another patient comes to mind. The young soldier had sustained a wound where a fragment of a mine had penetrated his upper thigh going through the femoral artery and vein leaving a small fistula between the two vessels. He was to be evacuated to a Center for vessel injuries. As he was being carried out of the hospital, he called out, and I examined him. The fistula had burst, and we rushed him to the OR immediately. I isolated the vessels above and below and with rubber compresses tight, the fistula was exposed, the vessels reconstructed, and the patient returned to the ward. At the last meeting of the Reunion of the 48th one of the nurses told me she knew the man and that he had no further trouble in civilian life.

One of my patients lives here in Inman, SC and recalls the Paris Hospital days. He insists that I saved his leg from amputation. These are the only two that I have heard from who were in the Wards during those hectic days.

All was not work during that winter when Germany refused to admit defeat. The Paris Opera had been told to produce only German operas during the occupation, so when freedom came they went wild with an explosion of the favorite French operas. A group of American Officers took the Royal box for the entire season and for about $2.50 per performance we enjoyed the magnificence of the Paris opera season.

Since I was the Unit's interpreter, I took it upon myself to take all the new officers to the Follies Bergere. Remember it was a cold winter with no heat in the living quarters, only the Wards. It snowed several times that winter. Baths were taken only once a week since the hot water was offered once a week. I can vividly recall how the goose pimples stood out on the skin of the beautiful nudes at the Follies. These were particularly noticeable when one used a strong pair of military field glasses. These evenings always ended in the late hours at Les Halles, the central market for Paris, where we ordered the unbeatable French onion soup, which was guaranteed to prevent both upset stomach and, or hangover.

The occupation money was almost a farce. Cigarettes were much better currency. I was sightseeing once over on the left bank and arrived at the Palais de Luxembourg. This had been the scene of much fighting for it was the Paris Headquarters of the Luftwaffe. There were numerous signs stating it was dangerous and to beware of mines. I came up to the sentry posted outside the gate and after sharing a cigarette, I started to depart. He asked if I would like to go through the

Palace? My reply was "no thanks" for I wanted nothing to do with mines. He laughed and said that these notices were just to keep the public out. I gave him the pack of cigarettes and toured the palace by myself. The marvelous paintings of Catherine de Medici by Titian had bullet holes in several, but on the whole the contents were intact. I was the only one in the building for a good two hours.

Our new colonel was a meticulous surgeon, using only mosquito hemostats, therefore I had the chance to pick up an entirely new way of operating. When he was transferred I heard that three hundred mosquito hemostats disappeared from the OR. He was good. He was the one who got us an invitation to watch the Chief Surgeon of the Red Cross Hospital in Paris operate. We arrived in time to enter the OR at nine o'clock. He stopped at noon. In the meantime he had done four major cases and three minor ones. He first did a partial gastrectomy (about half the stomach) in 45 minutes. He tied off the entire omentum in one tie. It looked like a sheaf of wheat. (Our colonel tied each vessel individually. I thought he would have a stroke). The French surgeon removed a large splinter from the thigh of a man. Then the French surgeon performed an hysterectomy, with an hemorroidectomy between the next case. He followed with another minor, a thyroidectomy was the next, another incision and drainage of an abscess, and as the last case a hernoirrhaphy was performed. We then adjourned for lunch. The meal lasted longer, over three hours, than did the surgery. I recall that the dessert wine was so good I asked the name three times, but I can't recall the name to this day. Must have been an exceptional wine. I did learn the use of a marvelously fast French needle similar to our aneurysm needle.

Being the interpreter for the unit had it's rewards. I was occasionally sent to the French celebrations as representative of the United States. One such occasion was a large banquet in the exclusive section of Paris to the left of the Etoile (Arc de Triumph) behind the buildings of the Trocadero. I was delivered to the party by jeep at eight o'clock, but the elite party goers did not begin to arrive until nine and, as most such parties, it went on and on. Now this was not too important to most of the guests, but the Metro stopped running at midnight. Sometime around one o'clock I gave my toast to "la Belle France" and United States and French unity. This was concluded by singing the Marseillaise. I went back to the apartment of my hostess where we tried to start a fire with a meagre amount of kindling. This was unsuccessful, so I judged it time to leave. I'm not at all sure, but I might have been supposed to spend the night. I was, and still am, very naive. At any rate I set out across Paris at about two a.m. Paris was still under blackout, and the walk to the Gare du Nord was a long one from the Arc de Triumph. There were still snipers in Paris during that time. I recall passing by the Parc Monceau during my promenade. This park is devoted to the Roman remains which have been found and reerected. The ghostly columns surrounded one part of the park, and they stood out in the dim light. I must admit that I had some fearful moments during that long trek.

Another of my assignments enabled me to meet the premiere pianist of France and attend her concert in the Salle Pleyell in the Trocadero Concert Hall. I was invited to the reception following, in her exquisite apartment which was also in the area of the Arc de Triumph. The windows overlooked the Seine with Napoleon's Tomb, and the Notre Dame Cathedral in the far distance. Doctor Silverberg, a dentist of our unit, had acquired the necessary invitation.

VE Day (Victory in Europe 8 May 1945) did not come any too soon for any of us. I had an operation scheduled for three o'clock that afternoon, but early that morning I persuaded the Colonel that the celebration was much too important to miss, and I would have plenty of time to catch the Metro back. I did not know what I was getting into. Getting to the Place de la Concorde in the center of Paris was no problem. Getting back was impossible. One would have to have been there to appreciate the wild gaiety of the natives on this day. It was almost impossible to move. The attempt to ride the Metro was a farce. I finally made it through the crowds, walking of course. Some kind soul covered for me, for the operation had been done long before my return.

Mr. Herbert Hucks, (at the time, Captain Hucks) was also in Paris at the time of the celebration of VE Day. He had been interrogating prisoners of war. He decided to view the celebration at the Arc de Triumphe since there was to be a ceremony of laying a wreath on the Tomb of the Unknown Soldier near the Eternal Flame at the Arch. When he arrived the crowd was already gathering. He decided to climb a lamp post in order to see the ceremonies better. He reached the crossbar and sat. Shortly afterward the crowd closed in. In a few minutes the entire SHEAF (Supreme Headquarters Expeditionary Armed Forces) contingent came through the crowd and stopped under and around our hero's lamp post. They had to wait for General de Gaule, who kept General Eisenhower, the Commander in Chief and numerous other big wigs waiting for thirty minutes until he deigned to arrive for the laying of the wreath. In the meanwhile, Captain Hucks dangled on the lamp post, hardly daring to breathe in the presence of so much "brass."

Not too long after the arrival of VE Day, (8 May 1945) orders came to assign me to a new hospital which had just arrived from the homeland and as was to be expected, was full of a bunch of eager beavers anxious to participate in concluding the war. It was ensconced in the magnificent quarters of Casserne Schleislig just outside of Metz. A casserne is a French army barracks, and ours was a permanent stone building, cold and a long way from being magnificent.

Two of my friends accompanied me in the jeep ride to the new assignment from Paris to Metz. On the way we stopped at the Verdun Monument where the shells had buried a whole company of Frenchmen as they stood erect, bayonets in place, ready to attack from the home trench. It is a pathetic tomb as the line of knives indicates the buried bodies beneath. We hurried through the magnificent town square of Nancy with the incomparable ironwork and fountains still standing but needing the rust removed and the golden gilding which adorns it today.

We stopped to enter the great hall of the Maginot Line, that marvelous Folly of the allies. Outside the railroad doors to the Fort Driant was a pile of dirty flat pancake metal discs about ten feet high. Two sergeants were sitting there working on one of the antitank mines, for such they were. When asked what the soldiers were doing, one replied he was taking it apart for the inside as a souvenir.

Our party thought they were crazy and moved quickly to put the shoulder of the Fort between us. A short time later we heard a loud bang, and after cautiously sticking our heads around the corner there was nothing to be found of the two sergeants or indeed of the pile of antitank mines, which had exploded a short time before. "The Englishman fights for the Empire, the Frenchman fights for the "Gloire de France," but the GI fights for souvenirs." We were repeatedly warned not to touch the numerous German destroyed material, for much of it was booby trapped with hand grenades.

Our hospital led a boring life for a few weeks around Metz, and again the soubriquet of "Old Ruins" was applied. I gave some lectures on the sights to be enjoyed around the old fortress, but my heart was not in it for I was ready to go home. The town was dusty, worn out with the war, and fed up with all soldiers.

So the word that we were to move was received as a mixed blessing for we were to be sent to Marseilles to await a ship to carry us to the Orient. Little did we know we were to have the pleasure of traveling by French rail. Not the Orient Express as depicted on TV, but the infamous Forty and Eight, i.e. forty men or eight horses in a freight car. For four days and nights we endured the hospitality of this method of transportation. Our gourmet food consisted of four choices of the Army famous K rations. Desert was always the emergency ration of the chocolate bar which was supposed to last three days. Elimination was a real problem, especially for the nurses. A moving train is no place to be when you "gotta go."

One of our numerous stops was in the old Roman town of Lyons. We were sidetracked to let a train pass which was full of Russian Troops being repatriated from Hitler's work camps. Neither they or we knew they were being sent back to the Soviet Union for execution or assignment to Siberia. They shouted and waved happily enough as they passed our train.

On the other side of our stopped train was a French train loaded with two large casks for each freight car. One of our curious troopers deducted the contents to be French wine. Soon the casks looked like porcupines as our boys stuck holes in the barrels and the wine spewed out. Happily the lads filled everything that would hold fluids with the precious liquor. They even emptied the water bottles and replaced it with wine. As our train pulled out the wine continued to spew out, and the French railroad men took the place of our troops in collecting it in everything available. The liquor relieved the boredom for a while, but unhappily nobody thought to tell our boys that this was green wine being shipped for storage for a year or more. Now green wine has the proper alcohol content for happiness, but other concomitant ingredients make for unhappiness. The result was a mighty sick group of American troops with as intransigent diarrhea, on a moving train without toilet facilities. Everybody was mighty glad when the port city of Marseilles appeared on the horizon.

We were put in an encampment above the city to await our boat to China. Entertainment was nil. On one dusty afternoon I stood in line with a metal canteen cup, to receive a Coca-Cola poured into the cup, without ice. Ahead of me stood four hundred other bored individuals with the same purpose in mind. No wonder that ex GI's refused to stand in line in civilian life.

Our CO was a rather peculiar individual who liked to tell the tale that some years in the past he had fallen (or was pushed) out of an airplane in the Western United States and had fallen on a slope of snow covered mountain, sliding down several hundred feet before being stopped by more snow. He was rescued, the broken bones reset, and rehabilitated. He and I had nothing in common, and frankly I thought him more than a little odd. At any rate when an order came through requiring me to return to my old unit the 48th General Hospital, I was more than happy to comply.

Again strings had been pulled and the old crowd had been assigned to a large hospital in the city of Marseilles. After joining the group, I was asked to take a jeep and go along the seashore road, the lower Corniche, to pick us out an officer's club from among the numerous villas lining the waters of the Mediterranean. We drove for miles along this famous playground beach, and all the lovely homes were marked "Keep Out - Reserved for the US Navy." When this was reported to the Colonel he arranged for us to enjoy the use of the home of the English Ambassador in the hills high above the city. It was a breathtaking home, built as a U with a Grecian colonnade surrounding the large swimming pool on the center, much like the eating center of the Metropolitan Museum in New York. However the Ambassador's pool had a sliding dance floor to cover the pool when it was not in use for bathing. There were numerous full size copies of Grecian statues throughout the building. Such opulence I have not seen before or since.

The town of Marseilles is not a pretty nor a happy town, then or now. A group of us were waiting on the side of the Rue Cannabierre, the main street, and as the bus stopped in front of the Hospital, someone stuck a knife in the back of the young Lieutenant standing beside us. Fortunately we could rush him into surgery immediately and he recovered uneventfully. Not a nice place at all. Marseille did make one attempt to entertain the troops stationed there. They put on a bullfight in the town arena. In France the bull is not killed in the end as is the custom in Spain. On the afternoon I went, the bullfight was rather tame and after the ceremonies ended the GI's went to the pens, captured the bulls, brought them into the arena and rode them in typical rodeo fashion, much to the delight of the American spectators and the chagrin of the French.

Some of us went fishing off the coast but after fishing all morning the total catch was only one three inch fish for the four of us. We ended up exploring the famous Chateau D'If outside the harbor. The story of the Man in the Iron Mask is based on the imprisonment of a member of the Royal family in the grim fortress. A swimmer nearby caught an eight foot octopus as we watched. The meat is tenderized by beating the tentacles on the rocks. The famed soup Bouillebaise, fish soup originated in Marseilles, and there it can stay so far as I'm concerned.

Five of us: Majors Buck Williams, Choctaw, "Ducky" Dawton, Kenneth Fairfax, and myself, were sent on Detached Service to the hospital in Nice, sometime in August 1945. All of us had been together since we met to form the 48th General Hospital. Now Detached Service is an Army euphemism for an extended vacation paid for by the taxpayers. We arrived at the assigned hospital in Cannes, which was a magnificent six story building, the top floor of which rotated to face the sun and was used as a solarium to treat the Tuberculosis patients. The whole front faced the Mediterranean Sea with a vista for miles. After signing in and leaving notice where we could be found if needed, the four of us headed for the Carlton Hotel on the front beach of Cannes. This hotel has twin towers on the front and the architect modeled the tower roof to duplicate the exquisite bosoms of his current girlfriend. The hotel was everything that superb hotels were supposed to be. We explored the region, going first up to the Victory Monument built by Augustus Caesar at la Turbie, above Monte Carlo, then to Grasse where the French perfume industry is located. We swam in the sea, but the South Carolina beaches put the ones of the Inland Sea to shame. They are composed of hard round pebbles that make walking difficult and lying down impossible. We finally settled for the opulence of Eden Roc, one of the world famous cafes along the shore. We never paid for anything, the hotels, the meals, the swimming were all paid by signing a "chit" of paper. Transportation on bus, train and "Metro" was all free. The swimming pool at Eden Roc was enhanced by Bikini clad models. I think this was the first time any of us had seen Bikini bathing suits other than in pictures for I'm sure none of the American girls would have been so brazen as to appear in public in such revealing outfits. I tried my best to have my friends take my picture with the models in the back ground but when the pictures were developed none of the girls appeared.

At the end of a week of this indulgence we were not in a mood to return to work, but the leaves could not be extended. Back we went to the 48th General in Marseille. Soon after our return, I received orders to report to England. It seems that all those of us with higher points were to be considered for return to the States. All along we had been collecting points based on length of service, Battle Stars (of which I had four), and a host of other requirements. Since I had the most in our unit, I suppose that was why I was chosen. My total at the time the Green Plan was announced was eighty-six. It seems that the States were beginning to experience a shortage of physicians, and those of us who had been away from home for two years or more were more than willing to return to alleviate that shortage. My orders were to report to the 216th General Hospital based near Cambridge, and originally from the Yale Medical School.

The transportation was from Marseille to Mildenhall Airbase in a DC-3, "the old workhorse." The pilot was a young Lieutenant, and since I was the only passenger, I was allowed to sit in the copilot seat. As we passed over the town of Nimes, the Roman aqueduct was spotted. Since I was still "old Ruins" I asked the pilot to go down closer. Which he did, much too close I thought. Then he asked if I would like him to fly through one of the arches. The need to investigate Roman remains evaporated quickly, and I assured him that I was not the least interested in flying through the arches. To this day I'm not sure whether he was teasing or not. A lot of those flyers were real nuts.

There was a genuine welcome to the 216th Hospital, for their main interest was in playing baseball. I sustained the only injury during the war playing ball! Somebody had driven a wooden peg into the ground at first base to hold the sandbag down, and my heel hit the peg instead of the bag. An x-ray showed no fracture but the bruise to the heel was incapacitating for some time.

There were many interesting sights to see around Cambridge, and the hospital was virtually at a standstill awaiting shipment back to the states. On one of our bicycle tours two of us ended up across the fens at Ely. The cathedral there was well worth the ride, and the home of Oliver Cromwell was visited also. On the back road of the Cathedral was a lovely garden. The gate happened to be half open so a Major friend and I cautiously set out to explore the flower filled area. Unfortunately at the end of the space was the Bishop of Ely who had been fast asleep in an easy chair set for him in a secluded nook. He welcomed the intrusive Americans just as if he had expected us. We enjoyed a chatty visit and a tea only to retreat in a somewhat chastened mood for having intruded upon the old gentleman, and for interrupting his nap.

It was during this stay outside Cambridge, that I received a second letter from the British War Office. This time was to notify me that I was to receive the Order of the British Empire (OBE). The Colonel agreed that henceforth I would have to be addressed as "Sir William," and arranged for me to report to General Eisenhower's headquarters with the letter. (I should have had a true copy certified, but in the excitement nobody thought about it.) When I arrived at the appropriate headquarters, two young lieutenants took the letter, and remarked "How the Hell did this get out of Headquarters?"

After disappearing for a short while, they returned to tell me that they would get in touch with me. Of course no one ever did. I am quite sure my old friend Sir Guy Campbell arranged the letter, for the name, rank and serial number and hospital designation were all correct.

An attempt was made to get to Edinburgh, but when I arrived at the train, it was so crowded that I would have had to stand all night, so I chose to visit the town of Sir William Harvey, who discovered the circulation of the blood. While there, in the pub I was induced to try Scotch. My remark was "I'd rather drink paint thinner. It tastes like the thinner smells." (So much for my education in the finer beverages.)

While I was down at the coast, the orders arrived for me to report to Grennock, for air transportation to the US. I jumped at the chance. Nobody told me that the Flying Fortresses returning to the States were being returned to be junked. We took off for Gander in Newfoundland, and halfway over, one of the engines stopped. I carefully stayed awake for the remainder of the night making sure that the remaining three engines continued their orderly functioning. All I recall about Gander is that it was snowing horizontally. The steel bucket seats of a converted bomber made for a cold cold flight, but I was happy to reach the ground, even covered with snow. (I refused to fly in any airplane until many years had passed.)

Fort Totten is at the end of Staten Island in New York, but not actually a part of New York. It was used during the Revolution to keep the British out of New York Harbor, but through the years has not been of very much military value. As soon as I arrived in the old military barracks, I looked up the Sergeant in charge of transportation. He evinced the dubious information that he could get me out of the town in a week or two. Since I had passed over six thousand miles in the previous two days, I allowed that better arrangements could be found. We agreed that he could cut the proper orders and I could be on my way on my own. He was to have the papers ready that afternoon, and in the meantime I could explore New York City.

It was my initial visit to the metropolis. Fifth Avenue was explored first. Since we had not had any ice cream for two years, the drug stores and ice cream parlors were targeted. I went from one side of the street to the other sampling the wares of each emporium. I really don't recall seeing much of New York, other than going out to the Statue of Liberty. Upon returning to the Fort surprise, surprise the papers were ready.

The following morning at the Railroad station I found a train ready to take me to the Promised Land. Getting a ticket was no trouble, finding somewhere to sit was no trouble, finding something to eat was no trouble, all this in marked contrast to the train travels in Europe, which were unbelievably crowded always, and there were no amenities to be found on the train coaches.

When I arrived at the Spartanburg Station there she was alone. Mother and Dad had kept little Bill to allow Gale to meet me. There were no flags, marching bands, and after enquiring of my war buddies, they had the same experience. No matter, it was enough to be home again. Also awaiting me was a promotion to Lieutenant Colonel, making me the Medical Officer with the highest rank in the area.

After a few days we drove down to the Processing Center at Fort Bragg, which was the nearest one that could release me from military service. The processing began in the morning, and there were not many to be processed out. When it became obvious that the personnel was not very aggressive about getting us separated that day, I pulled rank for the first time in my army career.

"Sergeant!" I said in a loud voice, "Under the Green Plan I have been rushed home from Europe and everything has been done to expedite my return until now. I have run one of these Processing Centers and I know how short a time it takes to get the job done. Now get me the officer in charge and I want to talk to him now!"

In a very short while, the remaining paper work was cleared, and we were on our way back to the normalcy of civilian life after five years of regimentation. I really believe that this is the reason that the armed services have difficulty keeping physicians in the ranks. Doctors are trained from the beginning of their careers to be independent in their thinking and in their actions and to rebel against the required regimentation even of the armed forces.

My bedding roll arrived from Europe in a very short time, and amazingly enough, everything was intact. I tried on the glasses which I had carried throughout the two years. I had acquired eight pair, since I am very nearsighted (20/400), and found that all had been ground, including the gas mask glasses, with the astigmatic correction vertically instead of horizontally, making them useless. As the French say, "Tant pis," or the GI's more graphic quote "SNAFU."

One of the most satisfactory elements of my return after two years overseas, was the fact that my young son no longer had to ask his mother whenever they were in a gathering where men in uniform were present, "Mother, is that my Daddy?" This had been a real emotional conversation stopper in the past.

And so ends that portion of the life history commonly referred to as the War Years. I would like to think the time spent under the nurture and tender mercies of the United States Armed Forces has been of some value both to me, to humanity and to the country itself. Heaven help us all if it proves to be otherwise!

THE END

MEDALS AND DECORATIONS

From the *Official Roster of South Carolina - World War II 1941 - 1946*: "Herbert, William C. Jr. O335916 Born: Clio, South Carolina (should be Timmonsville, South Carolina. Corrected by Mother) 12 June 14 Home Address: Spartanburg, Spartanburg County, South Carolina. Entered Active Duty Army 1 June 41. (First Lieutenant) Overseas European Theatre of Operations 2 June 44 to 23 September 45. AWARDS: ADSM (American Campaign) EAMECM (European African Middle Eastern Campaign with four battle stars) Northern France Ardennes, World War II medal National Defense Medal, Expert - C.W.S. (Chemical Warfare Service), Sharpshooter .45 Caliber Pistol. Honorable Separation Active Duty Lieutenant Colonel 1945."

Bartlet's Quotations

St. Lo.
Quote from a GI, "We liberated the Hell outa this town!"

EPILOGUE

In rereading the writings above, I seem to get a hint of that indefinable something which has so often baffled our enemies, be they British, German or Japanese. It has long been a mystery to others, as to how the American GI could take most miserable circumstances and make something amusing out of it, albeit, many times the humor was forced. This illustration comes to mind most often in the cartoons of World War II. In one, the General is sitting in a Jeep, watching a beat up G I Joe sweeping the road ahead with a Mine Detector and cheers up the soldier by saying "Take your time, soldier. Just do a good job." The Germans I met could never see the humor of the film "Stalag Seventeen," or the subsequent TV skits involving the misery of prisoner of war life. I hope in some small way I have been able to put across the leavening agent found by most Americans during this terrible infliction on the human race.

I do fear that in these writings I have not been able to make plain that we were upheld by the ingrained knowledge that we were indeed fighting to rid the world of one of history's most evil personages. Back of every hardship and danger was the shadow that if Hitler prevailed, the world would be precipitated into a Middle Ages from which we would never recover. I shudder to think how close civilization came to being wiped out, as I review the films of the possibility of the Nazi's acquisition of the atomic bomb. What would have been the consequences if the greatest amphibious landing the world had ever gathered together, had not succeeded?

How could we, as thinking human beings, have allowed the world to deteriorate to it's present level of instability as it presents in this year of 1993? How could we allow the rise of such second rate Hitlers as the Ayotola, Saddam Hussein, Kadaffy and petty tyrants? How could we, after devastating the whole world, with World War I, World

War II, Vietnam, and Korea allow such fiends, who deny every human right, to buy and make arms? Something has to stop this wholesale distribution of munitions which allows a few to dominate whole populations.

On a lighter note, I should like to see a series of World War II TV put together with the theme running through out, of the stupid orders that were issued during World War II, connivings and larceny of the procurement officers attached to each unit, be it Army, Navy, or Air Force. Their ingenuity cannot be equalled. The popularity of such a show should equal *M.A.S.H.*

Third Generation - The Born Officer.
William Chapman Herbert, III and Posey.

(TURN PAGE FOR MORE GOOD BOOKS)

______ HARRIET QUIMBY – $4.95
AN ACTIVITY BOOK FOR CHILDREN
Anita P. Davis / Ed. Y. Hall.
A companion activity work book for children (ages 10-14) to the book *Harriet Quimby – America's First Lady of the Air.* Soft Cover.

______ COMBAT SURGEON $12.95
Dr. William C. Herbert, Jr. M.D.
Memoirs of a U.S. Army Medical Corps Officer in Europe during World War II. Soft Cover.

______ DIVISION COMMANDER – $18.95
A BIOGRAPHY OF MAJOR GENERAL NORMAN D. COTA
Robert A. Miller.
An exciting biography of a true hero of World War II. A "fighting general" whose wartime record reads like a resume of the Mediterranean and European campaigns. Hard Cover.

TO ORDER

Please check the space next to the book(s) you want, send this order form together with your check or money order, include the price of the book(s) and add $1.75 for the first book and $1.00 for each additional book for handling and mailing to:

HONORIBUS PRESS
P.O. BOX 4872
SPARTANBURG, SC 29305

I have enclosed $ ________________ Check ____________________ or money order as payment in full. Please no COD's.

Name ____

Address ____

City ____

State ________________________________ Zip ____________

Please allow 2-3 weeks for delivery.